Planet Earth
Life in Danger

Planet Earth
Life in Danger

Babu Singh

Founder, Nature-Mankind Friendly Global Movement
Former Minister
Ward No. 1 Kathua
Jammu and Kashmir
India
e-mail: Savenature.mankind@yahoo.com

CBS Publishers & Distributors Pvt Ltd
New Delhi • Bengaluru • Pune • Kochi • Chennai

Planet Earth

Life in Danger

ISBN: 978-81-239-2186-0

Copyright © Author and Publishers

First Edition: 2012

Published by Satish Kumar Jain and produced by Vinod K. Jain for
CBS Publishers & Distributors Pvt Ltd
4819/XI Prahlad Street, 24 Ansari Road, Daryaganj,
New Delhi 110 002, India.   Website: www.cbspd.com
Ph: 23289259, 23266861, 23266867   e-mail: delhi@cbspd.com
Fax: 011-23243014   cbspubs@airtelmail.in

Branches

- **Bengaluru:** Seema House 2975, 17th Cross, K.R. Road, Banasankari 2nd Stage, Bengaluru 560 070, Karnataka
  Ph: +91-80-26771678/79   Fax: +91-80-26771680   e-mail: bangalore@cbspd.com

- **Pune:** Bhuruk Prestige, Sr. No. 52/12/2+1+3/2 Narhe, Haveli (Near Katraj-Dehu Road Bypass), Pune 411 041, Maharashtra
  Ph: 020-64704058, 64704059, 32392277 Fax: +91-020-24300160 e-mail: pune@cbspd.com

- **Kochi:** 36/14 Kalluvilakam, Lissie Hospital Road, Kochi 682 018, Kerala
  Ph: +91-484-4059061-65   Fax: +91-484-4059065   e-mail: cochin@cbspd.com

- **Chennai:** 20, West Park Road, Shenoy Nagar, Chennai 600 030, Tamil Nadu
  Ph: +91-44-26260666, 26208620 Fax: +91-44-45530020   email: chennai@cbspd.com

Printed at India Binding House, Noida, UP

to

Late Sh RP Saraf
my ideological–political teacher,
great social thinker,
a revolutionary leader
and founder of
Nature–Human Centric People Movement.

and

Late Sh Shamsher Singh Advocate
my father,
a man of simplicity, integrity
and full of human spirit,
who died untimely

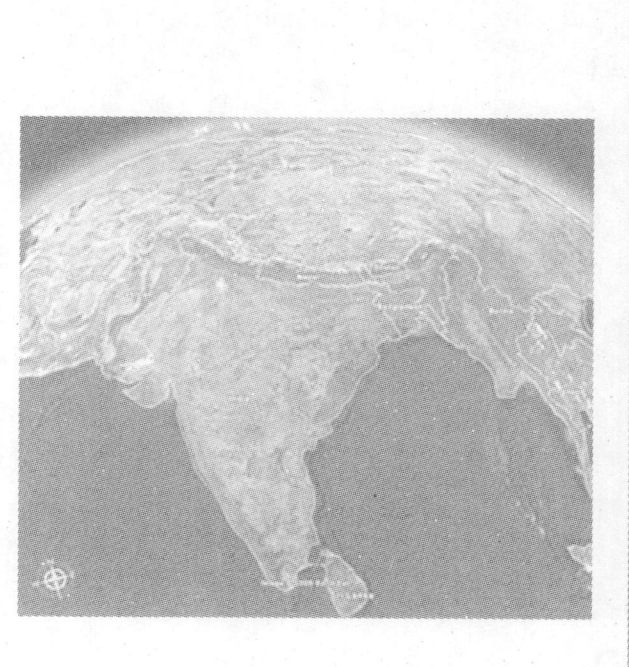

Preface

The book *Planet Earth: Life in Danger* presents detailed analysis (effects, causes and solutions) of global warming/climate change and global dehumanization/human cum social inequality, a fundamental existential challenge facing mankind today, which is threatening the very existence and survival of mankind and other bio-life on our planet Earth. The book is an attempt to survey, analyse and understand the reality of the present world at the start of third millennium and fundamental challenge confronting mankind today. The majority of prominent global scientists, climatologists, environmentalists and social thinkers are continuously giving warnings about the global danger of global warming/climate change from the last half century. There were so many international conferences before UNFCCC and 17 international conferences under UNFCCC held but nothing concrete had come out in these conferences. The unsustainable global corporate capitalist development model has unscientifically and ruthlessly degraded and damaged the environment of planet Earth and dehumanized the majority of mankind. The nature-mankind friendly philosophy guides us that the survival of human society is based on harmonious relationship between nature and mankind and within mankind. The nature and mankind are the two fundamental factors for the existence, sustainability and survival of human society. But the wrong and one-sided understanding of market regulated global corporate capitalist model and state regulated socialist-communist model about nature and mankind created big imbalance between nature and mankind and within mankind, resulted into a terrible global crisis of global warming/climate change and global dehumanization/human cum social-inequality. The approach of growth maximization and aim of profit/wealth maximization of global corporate capitalist and socialist-communist models oriented to misuse and abuse the natural and human resources and the large scale production of green house gases and impoverishment-dehumanization of global people, instead of maintaining focus, balance and harmony between nature and mankind and within mankind.

This book projects that the global corporate capitalist system operating through multiple nation-states (both the market regulated corporate capitalist model and state regulated socialist-communist model) is the fundamental cause of global warming/climate change and global dehumanization/human cum social-inequality. The immediate solution (short-term) of global warming/climate change and global dehumanization/human cum social inequality is to conclude legally binding save nature-mankind global treaty to cut globally the green house gases at that level which are necessary for the survival and sustainability of nature and mankind and to provide social-economic security to global people and to rationalize the global income differences. The fundamental solution (long-term) of global warming/climate change and global dehumanization/human cum social inequality is to fundamentally change the present global corporate capitalist system and to establish nature-mankind friendly global confederal new system.

The book thoroughly analyzes the historical process of global corporate capitalist system and explains the market regulated capitalist theory of Adam Smith and historical experience of capitalist model. It also explains the State regulated socialist-communist model of Marx and its historical experience throughout the world. The analysis of this book concludes that both the market regulated global corporate capitalist model and state regulated socialist-communist model are proved anti-nature anti-mankind and have become irrelevant and unworkable. The new fundamental global social reality, global challenge and global era need a new nature-mankind friendly global social model. In this book all the dimensions of alternative nature-mankind friendly global confederal new model are formulated comprehensively. The analysis of this book has been conducted in the light of nature-mankind friendly philosophy and nature-mankind friendly fundamental vision.

The writing of this book is based on my 36 years' deep and comparative study (including long political experience) of state regulated communist model, market regulated global corporate capitalist model, the model of international democracy and nature-human centric model under the guidance and long-close association with late Sh RP Saraf, a great scientific social thinker and the founder of underground proletariat (Marxist) party, internationalist democratic party and nature-human centric people's movement and editor of nature-human centric viewpoint. He was my political teacher and mentor, who always inspired and motivated me to work for the welfare of mankind, rational, sustainable and just advancement of global human society. The ministerial experience of seven portfolios in

Jammu & Kashmir (India) also helped me to deepen my global understanding about human society. I am with the core of my heart grateful to my all the activists and supporters who fully supported me to lay foundation of Nature-Mankind Friendly Global Movement to launch global movement against global warming/ climate change and global dehumanization/ human cum social inequality. They fully supported and encouraged me as they did always to write this book as a fundamental guiding model for mankind in general and missionaries and activists of Nature-Mankind Friendly Global Movement in particular.

I am very thankful to Mr Nishant who assisted me day and night to write this book. He helped in collecting information and data from internet, typing and editing process of the book.

I am extremely thankful to Prof. RK Ganjoo, Department of Geology, University of Jammu, J&K (India), who provided information with regard the study of Himalayan glaciers, different studies on climate change and fully cooperated in the editing process. I am also very grateful to Prof. GM Bhat, Department of Geology, Jammu University (India), who helped me during the process of editing. I am also grateful to my family, particularly my son Mr Samar who fully cooperated with me during the writing process of this book.

Babu Singh

Contents

1 Global Climate Change and Dehumanization/Human cum Social Inequality

Today mankind is confronted with a highly dangerous and fatal two-fold "environmental-humanitarian" fundamental existential challenge, the challenge of global warming/climate change and global dehumanization/human cum social-inequality. This is a very dangerous challenge facing mankind/ life and human civilizations on Planet Earth. If mankind (represented by various national govts of the world) will not give active and positive global response to the global challenge of global climate change and global dehumanization/human cum social inequality then the killer global disaster will parish the whole mankind/life and human civilizations on Planet Earth. We are living in a period when the process of change in human life, human society and human impact on the global environment has been accelerating at dizzying pace. It is now proceeding at a speed which puts the future of both the nature (environment of Planet Earth) and mankind at risks. The future of mankind and whole human civilization is not safe and sustainable because of degraded natural environment and global dehumanization.

In order to regularly assess the global threat of global warming/climate change the 'United Nations Environment Programme' and the 'World Meteorological Organization' established the intergovernmental panel on Climate Change as an organ of the United Nations in July 1986. We can trace the roots of UNIPCC in world earth day in 1970; the Stockholm conference in 1971–72; and the Villach conferences in 1980–1985. Under UNIPCC more than 10,000 scientists headed by Dr Pachori are working but Prof Bert Bolin (a meteorologist at

1

Stockholm University), Dr Robert Watson (an atmospheric chemist at NASA and now the chief scientist at the UK department of environment) and Dr John Houghton (an atmospheric radiation physicist at Oxford University) are the three chief ideologues of UNIPCC. Till now the UNIPCC issued four assessment reports about global warming/climate change. The first assessment report was issued in 1990, the second assessment report in 1996, the third assessment report in 2001, the fourth assessment report in 2007 and the fifth assessment report is likely to be published in 2013.

The Planet Earth in its historical process passed through many periodical existential natural disasters, in the form of global warming and global cool, (ice ages) due to the interplay of natural causes. But the majority of prominent international scientists under UNIPCC thinks that the present global challenge of global warming/global climate change is the outcome of human induced anthropogenic green house gases (GHG). But this book analyzed that the faulty and one-sided understanding and approach of global corporate capitalist system about nature and mankind and unsustainable corporate capitalist development model, are mainly responsible for the large scale emission of anthropogenic GHG and present fundamental global challenge of global warming/climate change and global dehumanization/human cum social inequality. The present crisis of global warming/global climate change has started from the origin of industrial formation or global corporate capitalism, which mainly focused on growth maximization and profit maximization. The industrial system or global corporate capitalism under both the models—market regulated corporate capitalist model and state regulated socialist/communist model unscientifically and ruthlessly exploited the natural resources and human resources, which has created great imbalance between nature and mankind and within mankind on Planet Earth. This imbalance has created a biggest threat for mankind and other bio-life in the form of global warming/climate change (extreme global warming and extreme global cooling) and global dehumanization, which can eliminate the human civilizations and life on Planet Earth. The prominent international scientists have continuously been giving warning to the world community about the danger of global warming/climate change. The world community under UNO conducted several international conferences and sixteen global climate conferences under United Nation Framework Convention on Climate Change (UNFCCC). In these conferences many positive decisions have been taken but till date the world community could not implement the Kyoto Protocol or could not finalize the legally binding global treaty regarding the reduction of GHG emissions. The time is running very fast and the ghost of global warming/climate change and global dehumanization is intensively haunting the whole mankind/life. But the mankind/world community under the global corporate capitalist system, which is operating through multiple nation-states, is lagging far behind from historically necessary goal to counter the threat of global warming/climate change and global dehumanization. The abovesaid global challenge is so serious that if it is not answered immediately by a nature-mankind friendly global model, the consequences can be very-very harmful and disastrous.

There are also skeptic views about cause, effect and solution of global warming/climate change. Some scientists believe that the global warming/climate change and global cooling is a periodical phenomenon or a natural cycle of Planet Earth. The Planet Earth experienced and passed through many times in global warming/global cooling and climate change in the past. The skeptic scientists having the opinion that the natural causes or variations in natural phenomenon are mainly responsible for global warming/climate change and the role of anthropogenic green house gases are insignificant. They contradicted the UN assessment reports prepared by UNIPCC about global warming/climate change. The skeptic scientists observed errors and falsehoods in the initial drafts of fourth assessment report of UNIPCC. The 'science and environmental policy project' set up a team B to prepare an independent evaluation of the available scientific evidences. The initial parallel organization took place at a meeting in Milan in 2003. It changed its name to Nongovernmental International Panel on Climate Change (NIPCC). The NIPCC was set up to examine the same climate data used by the United Nations sponsored

Intergovernmental Panel on Climate Change. The NIPCC organized an international climate workshop in April 2007. The NIPCC published a report under the title 'Climate Change Reconsidered' in the year 2009. This report stems from the Vienna workshop and subsequent research and contributions by a larger group of international scholars. The NIPCC is an international panel of nongovernmental scientists and scholars who have come together to understand the causes and consequences of climate change. The NIPCC scientists said that they are not predisposed to believe that the climate change is caused by the human induced green house gas emissions but they are able to look at evidences which the UNIPCC ignores.

In this book the comparative studies and analysis of UNIPCC and NIPCC (including the viewpoints of other skeptic scientists) about the threat of global warming/ climate change has been provided proper space. Both the studies of global warming/climate change—human centric and nature centric—greatly helped to conduct further study and research in order to develop a realistic understanding about the phenomenon of global warming/climate change. I got and utilized useful information and data from above-said international studies and the studies of other inter-national prominent social and natural scientists/thinkers. But I studied, investigated, analyzed and drew conclusions about the challenge of global warming/climate change in the light of nature-mankind friendly fundamental philo-sophy and vision. Similarly I studied and analyzed the social challenge of global dehumanization/human cum social inequality on the basis of comparative study of market regulated global corporate capitalist model and state regulated socialist/communist model, in the light of nature-mankind friendly fundamental philosophy and vision. In fact, the study of this book, 'Planet Earth-Life in Danger' considers that the threat/challenge of global warming/ climate change and global dehumanization/human cum social inequality is an integrated two-fold global challenge. The fundamental solution of this two-fold fundamental challenge is 'nature-mankind friendly global confederal' an alternative new social model.

CHALLENGE OF GLOBAL CLIMATE CHANGE

Today on Planet Earth the entire mankind is facing the biggest existential and deadly challenge of global warming/climate change, unheard in human history. The mankind never experienced such challenge in its existential history. This challenge is shaking our Planet Earth and endangering the entire mankind and other bio-life. This is a life and death challenge before human civilization. The challenge of global warming/climate change is not a new phenomenon but now it has grabbed the world's attention. The scientists—who track global environmental trends, such as rising temperatures, climate changes, ocean changes, melting of glaciers and ice sheets, deforestation, soil erosion and falling of water table— are saying that if these trends continue, the mankind will be in trouble. The Planet Earth's natural system slowly cycle from one Ice Age to next Ice Age. But the predicted rapid changes caused by the activities of mankind under unsustainable global corporate capitalist system are of deep concern. Many people will lose their lives as results of it. The destructive effects of higher temperatures are visible on many fronts. The crop withering heat waves have lowered grain harvests in key food producing regions in recent years. The soil erosion is lowering 30%, the inherent productivity of world's cropland. The *lesotho* and *mangolee* have reduced their grain production more than half over last three decades. Kazakhstan has abandoned its 40% grain land since 1980. The vast dust storms coming out of sub-Saharan Africa, Northern China, Western Mongolia and Central Asia remind us that the loss of top soil is not only continuing but expanding. The water tables are now falling in those countries that together contain half the world's population. The Saudi Arabia has announced that because its major aquifer has largely depleted, it will be phasing out its wheat production by 2016. The World Bank study shows that 175 million people in India are being fed by over pumping aquifers. In China this problem affects 130 million people. The climate change threatened food security, as after certain point, the rising temperatures reduce crop yields. The one degree Celsius rise in temperature above the norm, the farmers can expect 10% decline in wheat, rice and corn yields. The Planet Earth's average surface temperature has increased 0.6 degree Celsius or one degree Fahrenheit. The UNIPCC projected that the temperature will rise by up to 6 degrees Celsius (11 degrees Fahrenheit) during this century. The rising temperatures are reshaping the earth's ecology and geography. As the temperature of earth continues to rise, the mountain glaciers are melting throughout the world. The ice melts from glaciers in the Himalayas and on the Tibetan Plateau that sustain the major rivers of India and China. The irrigation systems depend on these rivers during dry season. In Asia both wheat and rice fields depend on the water of these rivers. The China is the world's leading wheat producer, the India number two and the US number three. The China and India also dominate world rice production. Whatever happens to the wheat and rice harvests in these two big countries (population giants), will affect food prices everywhere in the world. Rice growing river delta in Asia is threatened by the melting of these ice sheets. Even three feet rise would devastate the rice harvest in the Mekong delta, which produces more than half the rice in Vietnam, the world's number two rice exporter. The three feet rise in sea level would inundate half the rice land in Bangladesh, country of 160 million people. The fate of the hundreds of millions who depend on the harvests in the rice growing river deltas and floodplains of Asia is inextricably linked to the fate of these major ice sheets. The advancing of deserts (due to the result of overgrazing, over ploughing and deforestation) are encroaching on cropland in sub-Saharan Africa, the Middle East, Central Asia and China. The advancing of deserts in northern and western China has forced the complete or partial abandonment of some 24,000 villages and the cropland surrounding them. In Africa the Sahara is moving southward, engulfing cropland in Nigeria. It is also moving northward, invading wheat fields in Algeria and Morocco. The Somalia is an ecological disaster, with over population and overgrazing, that results in desertification, which is destroying its pastoral economy. In Latin America the deserts are expanding and forcing people to move in Brazil and Mexico. In Brazil 66 million hectares of land are affected. In Mexico, the degradation of cropland extends over 59 million hectares. The desert expansion and water shortage are now displacing millions of people. Each day the Mexicans risk their lives in the Arizona desert trying to seek jobs in US, which is herself suffering from unemployment due to economic recession. More than one lakh Mexican leave rural areas every year, abandoning their small plots of land. The conversion of cropland to other uses looms large in China, India and the US. The China with its massive industrial and residential construction and its paving of roads, highways and parking lots for a fast growing automobile fleet may be the number one in cropland loss. In the US the sub-urban sprawl is consuming large tracts of farmland. In India the villages are loosing the water from their irrigation wells to nearby cities. The China's farmers also loosing irrigation water to the country's fast growing cities. The food security will further deteriorate, unless the leading countries collectively mobilize to stabilize population, stabilize climate, stabilize aquifers, conserve soils, and protect cropland and to restrict the use of grain to produce fuel for cars. In the present world a dangerous geo-politics of food scarcity is emerging. Some countries are behaving with their narrow self interests, resulting negative trend. This trend started in 2007 when wheat exporting countries such as Russia and Argentina limited or banned exports in attempts to counter domestic food price rises. The Vietnam banned rice exports for several months for the same reason. The several other exporter countries also banned or restricted exports. This trend created panic in the scores of countries that import grain and forced them to negotiate food grain trade deal with food grain exporting countries. No country in the world is immune to the effects of tightening world food supplies, not even most developed countries. The food is the weak link, just as it was for many civilizations. The mankind is entering into a new food era, marked for higher food prices, rapidly growing numbers of hungry people and an intensifying competition for land and water resources, that has now crossed national boundaries as food importing countries try to buy or lease vast tracts of land in other countries. The land buying countries are mostly those whose populations have outburst their own land and water resources. The Libya importing 90% of its food grain and was one of the first to look abroad for land. The Libya reached an agreement with Ukraine to develop agriculture farm of 100,000 hectares of land in the Ukraine to grow food grain for its own people. The Saudi Arabia is looking to buy or lease agriculture land at least in 11 countries including Ethiopia, Turkey, Ukraine, Sudan, Kazakhstan, Philippines, Vietnam and Brazil. The Financial Times responded in March 2009 that the Saudis celebrated the arrival of the first shipment of rice produced on land they had acquired in Ethiopia, a country where the world food programme is currently working to feed 4.6 million people. Another major acquisition site for the Saudis and several other food grain importing countries is the Sudan. Ironically the Sudan is the place of the UN World Food Programme's largest famine relief effort. The Indonesia has agreed to give Saudi investors access to 2 million hectares of land to grow rice. The Chinese firm ZTE international has secured rights to 2.8 million hectares of land in the Democratic Republic of Congo to produce palm oil. Like Ethiopia and Sudan the Congo also depends on a UN World Food Programme. The China is also negotiating for 2 million hectares of land in Zambia to produce *jatropha*, an oil seed. The China has also acquired or has plans to acquire land in Australia, Russia, Brazil, Kazakhstan, Myanmar and Mozambique. The China signed an agreement with the Philippines govt to lease over one million hectares of land to produce crops. The South Korea a leading corn importer is a major investor in several countries and signed a deal for 690,000 hectares of land in the Sudan for growing wheat. The South Korea is one of the leading players in this food push. The world is seeing the early indicators, with record summer temperatures across the Europe in 2003 causing more than 20,000 deaths and Hurricane Katrina in 2005 killed more than 1800 in US. In addition to deaths directly attributable to climate change, there will be many more people who will lose their livelihoods, as farms become deserts and coastal communities will be submerged. The human civilization has evolved during a period of climate stability but this stability is drawing to a close. But we have entered into a period of rapid and often unpredictable climate change. Now the mankind has reached at a cross road where its survival and future is at stake. The condition of Planet Earth and mankind is very alarming. The gravity of this global challenge demands a correct and realistic understanding of its effects, causes and accordingly prompt global action for its solution.

The nature and mankind are two fundamental factors of human society. The nature and mankind constitute the social capital or supreme phenomenon of human society and play fundamental role in social development and progress. To give equal focus to both the nature and mankind is the basic need of balanced and sustainable development. The concept of nature is applied to whole universe while the term environment is only related to our Planet Earth. The mankind has social (collective) as well as individual nature (bio-social nature) which is divided among different social units, groups, classes and formations. The human society change and develop through two-sided interaction between nature and mankind on the one hand and among the different social units, groups, classes and formations of mankind through the alternate motion of unity and struggle on the other hand. The reality is that the mankind is the part of nature not the master. The mode of existence of mankind is biological (individual) while its mode of living is social. This characterises human being as a bio-social phenomenon (individual cum social). The existence and balanced development of human society is based on harmonious relationship between nature and mankind and within mankind.

The unrealistic approach towards nature and mankind has created a biggest challenge before mankind. The approach of infinite nature and human centric vision has disturbed the harmonious relationship between nature and mankind. All ancient theories, including the theory of claiming supernatural origin, speak of a divinely created limitless existence with mankind as its supreme creation. In 19th century Karl Marx too holds Nature as inexhaustible (having no beginning and no end) and the people are the makers of the history. The western liberal schools of thought of 19th and 20th centuries are also one in stressing mankind as the unique phenomenon of nature. According to scientific facts the human approach emphasising the boundless character of reality and primary position of human being in the system of creation is merely an unrealistic view. The nature with regard to Planet Earth is limited in matter, space and time. It is characterized by the interdependence and two-sided interaction among and inside the different natural processes (including the process of mankind) through the process of alternate unity and struggle.

The unrealistic approach was bound to lead towards distortions in human thinking, practice, organization and behaviour. The entire historical process of human society in the form of various social stages—food gathering stage, clan stage, feudal monarchical stage, industrial national capitalist stage and now the global corporate capitalist stage —shows that the human society up to the feudal monarchical age has marched mainly with the tune of nature and the exploitation of human resources has started from the clan stage. But after the emergence of industrial technology and corporate capitalism (both national and global corporate capitalism from AD 1750 up to now) the human society has one-sidedly advanced the developmental process at the cost of nature (by degrading environment of Planet Earth) and mankind (by misusing human resources). We can also trace the genesis of our current dilemma of global warming in enormous growth of unsustainable enterprises since 1900 AD, the 20-fold expansion of world economy and fourfold increase of world population. The UN IPCC report states that the Planet Earth was pollution free before the emergence of Industrial Age or corporate capitalism. In the year AD 1750 pollution began to appear on Planet Earth due to the use of unsustainable industrial technology, unsustainable raw materials and polluting energy. The process of environmental degradation has also started after the emergence of unsustainable industrial technology, unsustainable raw materials, polluting energy and corporate capitalist system with the production of anthropogenic green house gases. Up to the 19th century the unrealistic approach towards nature and mankind could not cause any major harm to the natural system of Planet Earth and mankind. The mankind at that time knew too little to exploit the different elements of nature and human resources but as soon as the human knowledge of nature began to advance, the mankind particularly of developed world started to apply its knowledge to overuse and abuse the nature through science and modern technology. But now the global corporate capitalists or the rulers of the world are at large and global scale overusing, misusing and abusing the natural and human resources in both developed and developing world in order to attain maximum growth and maximum profit. The unprecedented increase of human population also degrades the environment of Planet Earth, depleting natural resources and impoverishing majority of human community. Today the world population numbers over 700 billion. Over 90 million people are being added every year; by this speed the global population will reach more than 10 billion by the year 2050. The earth's carrying capacity is believed to be 8 billion people. If the present population rate continues then there will remain very less space on Planet Earth for mankind to live. The large scale exploitation of natural resources and human resources along with the pressure of increasing population resulted into environmental degradation and dehumanization of humanity manifold, which is posing a global threat to the survival of mankind. This global threat is the outcome of one-sided and wrong philosophical approach about the relation of nature and mankind. The industrial corporate capitalist system (both market regulated corporate capitalist model and state regulated socialist-communist model) now advanced from national stage to global corporate capitalist stage with the aim of attainment of maximum growth and wealth maximization. The global corporate capitalist system adopted anti-nature and anti-mankind unsustainable development model in which the polluting technology, unsustainable raw materials and polluting energy have damaged the environment of Planet Earth ruthlessly and the aim of wealth maximization created crude global dehumanization/human inequality. The blind use of polluting technology, unsustainable raw materials and polluting energy in the sphere of industrial manufacturing production sector, chemical based inorganic agriculture sector, infrastructure sector and service sector (transport sector, aviation sector, shipping sector, information technology sector, fossil fuel energy sector, thermal energy sector, etc.) are immensely damaging the global environment of Planet Earth. The greed of "maximum growth" and "maximum wealth" has expanded the polluting science and technology, polluting energy, use of polluting and unsustainable raw materials, unsustainable corporate capitalist development model, misuse and overuse of natural and human resources at global level , resulted into large scale production of green house gases like carbon dioxide, methane, nitrous oxide, chlorofluorocarbon [CFC], etc. The unsustainable global corporate capitalist development model and the large scale production of green house gases or heat trapping gases in the world has damaged, degraded and imbalanced the environment and natural climate system of Planet Earth and mankind. This imbalance has created a fatal crisis of global climate change and global dehumanization/ human cum social inequality. If the global corporate capitalist system will continue with overloading the atmosphere with green house gases, deforestation, polluting and disturbing oceans with pollutant wastes, oil spills, drilling of oil in deep seas, overfishing and glacier melt fresh water which dilute the salinity, over pumping of underground water and aquifers, overgrazing, over ploughing, massive exploitation of natural resources and human resources, than we will confirm the death warrant and final doom of whole mankind including other bio-life and human civilization.

The majority scientists of the world under UNIPCC believe that the present challenge of global climate change is mainly caused by the human induced anthropogenic GHG. But some scientists believe that the anthropogenic GHG are not the main or alone factor responsible for present global climate change. They uphold that the global warming is a complex periodical natural cycle and multi-natural factors are responsible for global climate change. According to them the geological evidences indicate that the phenomenon of climate cycle (glacial and inter-glacial period) or periodical global warming and global cooling happened in the past due to the interplay of natural causes since the origin of the Planet Earth. In the geological past there have been significant episodes of global warming and global cooling not associated with human activities. The phenomenon of global warming/climate change is not a function of human induced carbon dioxide in atmosphere alone but of the mechanism beyond human interference.

The nature-mankind friendly global confederal model uphold that (along with the human induced factors) the natural factors such as (1) variation in solar activity (2) orbital variations of the earth (3) volcanic activities (4) ocean variability (5) plate tectonics (6) changes in atmosphere and damage of ozone layer and (7) changes in electro-magnetic field, should also be seriously observed, studied and regularly assessed that how much they are inducing the present phenomenon of global warming/climate change as it happened in the past of Planet Earth.

All the indicators of nature's degradation (environmental degradation of Planet Earth) are sounding the loud signal and majority of prominent international scientists and environment experts are continuously giving serious warning from so many years about the global warming/climate change. The situation is very alarming and it will be worsen within years if a global united action is not taken immediately. There is a historical need to sign a global treaty named as Save Nature: Mankind Global Treaty in order to effectively tackle the global warming-climate threat. If the mankind (which is living in different national systems within the framework of global corporate capitalist system) will not respond positively and seriously to the threat of global warming/climate change and global dehumanization/human inequality immediately, then be ready to face the reaction of the global people who are the victim of global warming/climate change and dehumanization and the nature's fury and ever biggest global disaster never seen in human history, which will perish and destroy the whole mankind and other bio-life on Planet Earth. This is an earth shaking and unprecedented global existential challenge. The existence of life or non-existence of life or continuity of human civilization is a question mark before mankind. The life on Planet Earth is in danger. To save nature-mankind from global disaster of global warming/climate change and global dehumanization/human-inequality through nature-mankind friendly global response is the topmost global priority of mankind.

GENESIS AND EFFECTS OF GREEN HOUSE GASES

The green house effect is the heating of the surface of planet or moon due to the presence of an atmosphere containing gases that absorbs and emits infra-red radiations. Thus green house gases trap heat within the surface/troposphere system. This mechanism is different from actual green house, which works by the isolating warm air inside the structure so that heat is not lost by convection. The term "green house effect" originally came from the green house used for gardening and vegetable production. The green house effect was discovered by Joseph Fourier in 1824, firstly experimented by John Tyndoll in 1858 and reported by Arrhenious in 1896. The earth receives energy from the Sun mostly in the form of visible light (UV rays, infra-red rays, etc.). The 50% of this energy is absorbed at the earth's surface. The earth radiates energy in the infra-red range. The green house gases in the atmosphere absorb most of the infra-red radiations emitted by the surface and pass the heat to the other atmospheric gases through molecular collision. The green house gases also radiates in infra red range. The radiation is emitted both upwards (with part escaping to space) and downwards to the earth's surface. The surface and lower atmosphere are warmed by the part of energy that is radiated downwards.

The global warming of the earth surface and lower atmosphere is the result of an enhanced green house effect. This happened due to increase in atmospheric green house gases produced by human activities. The part of human produced gases is called anthropogenic green house gases which are inducing global warming.

The main green house gases in the atmosphere are water vapours, carbon dioxide, methane, nitrous oxide, CFC, ozone, etc. The green house gases greatly affect the temperature of the Planet Earth. The water vapour contributes 36–72%, carbon dioxide 9–36%, methane 4–9%, ozone 5–7 %. The other green house gases include sulphur hexafluoride, hydrofluoric carbons and per fluorocarbons. The nitrogen tri-fluoride has high global warming potential but it is present in small quantity. The earth's surface without green house gases would be an average 33 degrees centigrade colder than present temperature. The green house gases absorb and emit radiations within the infra red range. This process is the fundamental cause of green house effect. The human activities since the beginning of Industrial Era (from AD 1750) increased the levels of green house gases in the atmosphere. The UNIPCC 2007 assessment report noted that the atmospheric concentration of green house gases, aerosols, land cover and solar radiations altered the energy balance of climate system. The increase in anthropogenic gases concentration is the cause of most of the increase in global average temperature since the mid-20th century. Since the start of Industrial Era the human activities had added green house gases to the atmosphere mainly through the burning of fossil fuels and damaging of forest cover (deforestation). Many experts estimated that the average temperature will rise 1.8 to 5.8 degrees centigrade by 2100. The rate of temperature increase would be much larger than most past rates of increase. The majority of climatologists concluded that the human activities are responsible for most of the global warming/climate change. The human activities contribute to global warming/climate change by enhancing earth's natural green house effect. The main human activities that contributed to the global warming/ climate change are use of fossil fuels; (coal and petroleum products), chemical based inorganic farming, unsustainable industrialisation, nuclear energy and manufacturing of nuclear bombs, large scale mining of Planet Earth, use of information technology globally, drilling of oil in the oceans, unbalancing of oceans due to glaciers and Arctic ice melting and large scale deforestation of Planet Earth. The most of the burning of fossil fuels occurs in the automobiles, factories and in thermal power plants. The burning of fossil fuel creates carbon dioxide and chemical fertilizers create methane and nitrous oxide. The most of the production of green house gases comes from the unsustainable global corporate capitalist development model, extravagant and wasteful lifestyle of developed rich countries and big developing countries. The US and China produces 50% green house gases and 30 developed countries emitting 80 percent green house gases in the world.

EFFECTS OF GLOBAL CLIMATE CHANGE

The global climate change is disturbing environment and climate system of Planet Earth. The effects of rising temperature are very dangerous and destructive. The higher temperature diminish crop yields, melt the mountain glaciers that feed rivers, generate more destructive storms, increase the severity of flooding, intensify droughts, cause more frequent and destructive wild fires and alter ecological system everywhere.

We can anticipate extreme weather events due to warmer climate. The global warming is continuously melting Arctic ice cap (north pole ice cap) along with Greenland. Mark Serreze, a veteran Arctic specialist said, If you asked me a couple of years ago that, when the Arctic could lose all of its ice, then I would have said 2100 or 2070. The Guardian reported that the Greenland ice cap is melting so fast that it is triggering minor earthquakes as pieces of ice weighing several billion tons each break off the ice sheet and slide into the sea. But now I think that 2030 is a reasonable estimate, Robert Corell, chairman of the Arctic Climate Impact Assessment reported that we have seen a massive acceleration of the speed with which these glaciers are moving fast into the sea. The ice is moving at 2 meters an hour on front 5 km long and 1500 meters deep. The Arctic region is experiencing some of the most rapid and severe climate changes on Planet Earth. The Arctic scientist Julienne Strove observed that the shrinking Arctic sea ice may have reached, a tipping point that could trigger a cascade of climate change reaching into earth's temperate regions. The warming of the Arctic is a defining event in the history of Planet Earth. The Arctic Climate Impact Assessment report noted that the retreat of the sea ice has devastating consequences for polar bears, whose very survival may be at stake. The Arctic Climate Impact Assessment team, an international group of 300 scientists conducted the study and found that in the regions surrounding the Arctic including Alaska, Western Canada and Eastern Russia, the winter temperature have climbed by 3.3 degrees Celsius over the last half century. Chris Ripley, head of the British Antarctic survey said, "the ice is moving faster both in Greenland and in Antarctic, than the glaciologists had believed". We can see from ice melting that human civilization is in trouble. If Greenland ice sheet melts the sea level will rise 7 meters. If the west Antarctic ice sheet breaks up and many scientists think it could go before Greenland and will add another 5 meters to the increase. The glaciers which are the main source of fresh water and rivers are melting and retreating. The total surface area of glaciers worldwide has decreased manifold by the end of 19th century. The current glaciers retreat rates and mass balance losses have been increasing in the Andes, Alps, Pyrenees, Himalayas, Rocky mountain and north Cascades. The Himalayan glaciers that are the biggest source of Asia's biggest rivers—Ganges, Indus, Brahmaputra, Yangtze, Mekong, Salween and Yellow river could disappear as temperature rises. The 2.4 billion people live in the drainage basin of the Himalayan rivers. The India, China, Nepal, Pakistan, Bangladesh and Myanmar could experience unprecedented rains and floods followed by droughts in coming decades. In India, the Ganges alone provides water for drinking and farming to more than 500 million people.

The Asian cities are endangered by climate change. The Asia's coastal mega cities will flood more on a large scale and affect million more people, if current climate change trend continue. The report of October 2010—climate risks and adaptation in Asian coastal mega cities, examines the impact of climate change on Bangkok, Ho Chi Minh city and Manila, under a range of different scenarios. The report is the product of a two-year collaborative study by the Asian Development Bank, the Japan International Co-operation Agency and the World Bank. It was released at the Asia Pacific Climate Change Adaptation Forum. The report find that the costs from major flooding events on infrastructure and the economy could run into the billions of dollars, with urban poor populations likely to be the worst hit. All these coastal mega cities will face climate related risks, such as rising sea levels and frequency of extreme weather events. The report also recommends that the govt/administration of coastal mega cities should undertake pro-active measures to address climate risks as an integral part of urban planning. In Bangkok flooding is caused by land subsidence and increased rainfall in the large watershed that drains through city. In Ho Chi Minh city the report states that around 26% of the population is currently affected by extreme storm events but those numbers could climb to more than 60% by 2050. The report states that the main threats to Manila are extreme rainfall, sea level rise, as well as more powerful typhoons. The latest survey of 170 nations compiled by British based global risks advisory firm Maple Croft published in October 2010, ranks South Asia as world's most climate vulnerable region, its fast growing population badly exposed to floods, droughts, storms and sea level rise.

According to a survey, 16 listed countries being at extreme risks from climate change over the 30 years, five are from South Asia with Bangladesh and India in first and second places, Nepal in fourth, Afghanistan in eighth and Pakistan at sixteenth place.

The rise of green house gases trigger an unprecedented rate of global warming which will result in the loss of the ice covered polar seas by 2020. The melting of polar ice caps and glaciers of Planet Earth will raise the sea level, resulting submergence of many countries and islands (Bangladesh, Maldives and Tuvalu) and cities (e.g. New York, London, Amsterdam, Mumbai). The global warming/climate change is raising the sea level. The current sea level rise has occurred at a mean rate of 1.8 mm per year for the past century. The increasing temperature results in sea level rise by the thermal expansion of water and through the addition of water to the oceans from the melting of continental ice sheets. The thermal expansion is the primary contributor to the sea level rise. The UNIPCC fourth assessment report (2007) predicted that by 2100, the global warming will lead to a sea level rise of 19 cm to 58 cm, depending on which of six possible world scenarios comes to pass. These sea level rises could lead to difficulties for shore-based communities in the next centuries. The major cities such as London, New Orleans, etc. already need storm-surge defences and would need more if sea level rose, though they also face issues such as sinking of land. The sea level rise could also displace many shore-based people. It is estimated that the sea level rise of just 20 cm could create 740,000 homeless people in Maldives, Tuvalu, Nigeria, Bangladesh and other low-lying countries. These countries are among the areas that are at the highest level of risk. The UN environmental panel has warned that, at current rates the sea level would be high enough to make the Maldives uninhabitable by 2100. The future sea level rise, like the recent rise, is not expected to be globally uniform. The sea-level rise in some regions shows substantially more than the global average (in many cases of more than twice the average) and other regions the sea level fall.

It is well-known fact that the glaciers are subject to surges in their rate of movement with consequent melting when they reach lower altitudes or the sea. The 2004 Arctic Climate Impact

Assessment Climate Models projected that the local warming in greenland will exceed 3 degrees centigrade during this century. The ice sheet models projected that the global warming would initiate the long-term melting of the ice sheet, leading to global sea level rise. The study of environment and urbanization [April 2007] reported that 634 million people live in coastal areas within 30 feet of sea level. The study also reported that about two/third of the world's cities with over five million people are located in these low-lying coastal areas. The UN IPCC report of 2007 estimated that the accelerated melting of the Himalayan ice caps and the resulting rise in sea levels would likely increase the severity of flooding in the short-term during the rainy season and greatly magnify the impact of tidal storm surges during the cyclone season. The sea-level rise of just 40 cm in the Bay of Bengal would put 11% of the Bangladesh's coastal land underwater, creating 7 million to 10 million climate refugees. The International Institute for Environment and Development study pointing out that 634 million people currently live along coasts and rice growing river deltas. One of the most vulnerable countries is China with 144 million potential climate refugees. India and Bangladesh are next with 63 million and 62 million, Vietnam 43 million, Indonesia 42 million, Japan with 30 million, Egypt with 26 million and USA 23 million vulnerable people. The Bernard Francou, research director for the French govt. Institute of Research and Development believes that 80% of South American glaciers could disappear within the next decade. The Peru the site of 70% of the earth's tropical glaciers is in trouble. The 22% of its glacial endowment which feeds many Peruvian rivers that supply water to the cities in the semi-arid coastal regions has disappeared. Mr Wilfred Haiberli head of the World Glacier Monitoring Service reported in 2009 that some 90% of the glacial ice in Spain's Pyrenees mountains has disappeared over the last century. Daniel Fagre US geological survey ecologist at glacier national park reported in 2009 that the park's glacier, which had been projected to disappear by 2030, may in fact go by 2020. The UNIPCC assessment reports suggest that the deltas and small island states are particularly vulnerable to sea level rise caused by both thermal expansion and ocean volume. The relative sea level rise is currently causing substantial loss of lands in some deltas. The UNIPCC have found that this is a serious risk scenario in coming decades. The Polynesian islands of Tuvalu may also sink due to sea level rise. John Hunter in his study reported that Tuvalu had been experiencing sea-level rise of about 1.2 mm per year. Many countries of the world are loosing major portion of their costal areas or most of their land mass and many islands will be drowned entirely like Louisiana, Florida, etc. in US, Northern region of Canada, part of Russia, part of Gujarat, Mumbai, and East South costal line of India, Bangladesh, part of the Netherlands, islands of South East Asia, part of Europe, part of Mexico and Brazil, Maldives, part of Australia, Caribbean region, etc.

The atmospheric brown clouds, which consist of soot particles from burning of coal, fossil fuels and wood are affecting seriously on global climate. The rise in evaporation due to global warming/climate change will cause heavy rainfall and devastating floods with more erosion in entire Planet Earth. The erosion, in turn, can be invulnerable to tropical areas, leads to desertification, especially in Africa. Andrew Goudie Professor of Geography at Oxford University reports that the number of Saharan dust storms has increased 10 fold during the last century. The most affected African countries by soil loss from wind erosion are Niger, Chad, Mauritania, northern Nigeria and Burkina faso. Nearly three billion tons of fine soil particles that leave Africa each year in dust storms are slowly draining the continent of its fertility and biological productivity. The dust storms leaving Africa travel to westward across the Atlantic, depositing so much dust in the Caribbean where they cloud the water and damage coral reefs. The people in China are also facing dust storms that originate in the north-west China and Mongolia. But the rest of the world should learn about this fast growing ecological catastrophe when the massive soil-laden storms leave the region. The water erosion also takes toll on soils. This can be seen in the silting of reservoirs and mud/silt laden rivers flowing into the sea. Pakistan's two large reservoirs, Mangla and Tarbela which store Indus river water for Pakistan's vast irrigation network, are losing roughly one percent of their storage capacity each year as they fill with silt from deforested watersheds. The Ethiopia (a mountainous country) is losing near about two billion tons of top soil in a year, washed away by rain. This is one of the important reason that Ethiopia always seems to be on the verge of famine and never able to provide meaningful food security. The soil erosion from the deterioration of grasslands is widespread. In the world near about 200 million people make their living as pastoralists, tending cattle, sheep and goats. Due to overgrazing half of the world's grasslands have degraded. This problem is experienced in Africa, Middle East, Central Asia and north-west China. Nigeria (Africa's most populous country) is losing 351,000 hectares of range land and cropland to desertification each year. In the South-Eastern Province of Sisten Baluchistan sand storms, have buried 124 villages and forced the people to abandon. The UN environment programme team reported that in Afghanistan up to 100 villages have been submerged by wind blown dust and sand. The agriculture land has also been seriously affected in Afghanistan. The China's cattle, sheep and goat populations spiralled upward. The China's grazing capacity is 82 million cattle, while the US grazing capacity is 97 million cattle. But the US has only 9 million sheep and goats and China has 284 million. The sheep and goats are destroying China's western and northern provinces land's protective vegetation and converting productive range land into desert. The China's desertification may be the worst in the world. It is not invading armies that are claiming China's territory but expanding deserts. The Wang Tao world's leading desert expert reported that by the end of century nearly 3600 sq km are going to be desert annually. He reported that over the last half century 24,000 villages in northern and western China have been entirely or partially abandoned as a result of being overrun by drifting sand. The mankind will further face dangerous and catastrophic cyclones, floods, droughts and scarcity of drinking water. Historically the agriculture began more than 10,000 years ago and crops have been developed to maximize yields in a relatively stable climatic regime. Now that stable climatic regime has disturbed and rapidly changed, the effect is likely to be on the crop yield. Since the crops are grown at or near their thermal optimum, even a relatively minor increase in temperature 1 or 2 degrees Celsius during the growing season can shrink the grain harvest in major food producing regions, such as North China Plain, the Gangetic Plain of India and the US Corn Belt. The higher temperatures can halt photosynthesis, prevent pollination and lead to crop dehydration. The scientists of India Mr KS Kavi Kumar and Jyoti Parikh assessed the effects of higher temperatures on wheat and rice yields. Their model is based on data of ten sites wherein they concluded that in north India one degree Celsius rise in mean temperature did not meaningfully reduce wheat yield but two degrees rise lowered yields at almost all sites. The global climate model shows that as temperature rises, some parts of the world will become more vulnerable to droughts. Heat plus droughts can prove deadly in south western US and Sahelian region of Africa. The Sahel region that stretches across Africa from Mauritania and Senegal in the west to Sudan, Ethiopia and Somalia in the east is already suffering from devastating periodic droughts and high temperatures. Now the low rainfall in this region is becoming even sparser.

The irrigation and hydel energy system of the world will collapse. The Earth's underground water table will be lowered down. The crisis of falling water table and the shrinkage of irrigated agriculture are more dramatic in Saudi Arabia. The Saudi Arabia developed a heavily subsidized irrigated agriculture largely based on pumping water from deep fossil aquifer in order to become self-sufficient in wheat production. Now Saudi Arabia announced that their aquifers are largely depleted and they would reduce their wheat planting by 1/8th each year until wheat production ends by 2016. It is not the Saudi Arabia alone but the scores of countries who are over pumping aquifers to meet their growing water needs. The farmers, who lose their irrigation water, have the option of returning to dry land low yield farming that depends on rainfall. But in more arid regions, such as in south-western US and parts of middle east, the loss of irrigation means the end of agriculture. In Yemen the water table is falling by roughly six feet in a year. The falling water table are already adversely affecting harvests in some larger countries, including China, India, and US. The water shortage and falling water table are more serious in India. Till date 100 million farmers of India drilled more than 21 million wells and invested approximately 12 billion dollars in wells and pumps. The report of new scientist stated that the half of India's traditional hand dug wells and millions of shallow tube wells have already dried up, forcing the farmers to migrate. The growth in India's grain harvest squeezed both by water scarcity and the loss of cropland to non-farm uses. The World Bank study report in the year 2005 stated that 15% of India's food supply is produced by mining ground water and 175 million Indians are fed with grain produced by water mining which is depleting. In the US, the USDA report stated that, in parts of Texas, Oklahoma and Kansas, three leading grain producing states, the underground water table has dropped by more than 30 meters. As a result, wells have gone dry on thousands of farms in the southern great plains, forcing the farmers to return to low-yielding dry land farming. Pakistan, with 177 million people, is also mining underground water. In the fertile Punjab plain the falling of water table appears to be similar to that in India. The observation wells near Islamabad and Rawalpindi show a fall in the water table from 1 to 2 meters in a year. In the province of Baluchistan the water tables are falling by 3.5 meters per year, indicating the day when the city will run out of water. Iran is also over pumping its aquifers by an average of five billion tons of water per year. It too will face a day of reckoning. Israel is also depleting both of its aquifers, the coastal aquifers and the mountain aquifers. The over pumping of aquifers is occurring in many countries. The lack of stable supplies of water is reaching at critical proportions particularly for agriculture purpose and the problem will further worsen because of rapid urbanization worldwide. Today experts assessed that about 21 countries with a combined population about 600 million to be either cropland scarce or fresh water scarce. The depletion of aquifers in the world means creating potentially unmanageable food scarcity. The global agriculture system and global fishing sector will be severely affected. In Latin America the deserts are expanding and forcing people to move to Brazil.

The entire mankind will become the victim of hunger and fatal diseases throughout the globe. The global network of satellite communication, telephone and mobile services, global air services, spacecraft network, space station, lift system of institutions and buildings, metro rail services of the world, global shipping communication, global internet services, etc. will collapse due to the effects of global climate change.

The global warming/climate change is seriously damaging and imbalancing the global ocean currents or thermo-haline ocean circulation and marine life, which is the biggest factor for the survival of mankind and bio-life of Planet Earth. The changes in ocean circulation and shut down of ocean thermo-haline cycle or the great ocean conveyor belt will parish the whole life [human, marine and bio-life] of Planet Earth. The oceans are the great source of fresh oxygen [bigger than global forest cover] and serve as the biggest sink for carbon dioxide and taking up much that would otherwise remain in the atmosphere. The increased level of carbon dioxide has led to ocean acidification. When the temperature of the ocean increases, they become less able to absorb the carbon dioxide. The climate of Planet Earth is mainly governed by the oceans. The ocean circulation conveyor belt or the ocean thermo-haline circulation helps to balance the climate of Planet Earth. The oceans surface currents redistribute heat around the globe and have profound effect on the global climate. The gulf stream and the north Atlantic currents carry huge volumes of warm salty tropical water from north to the Greenland coast and to the Nordic seas. Many scientists are warning that the north Atlantic might cool down by the turn of this century due to global warming/climate change. The global warming may trigger the events that could not only slow the supply of tropical water flowing towards North, but it could also disrupt the entire ocean circulation pattern. The impact of ocean circulation on global climate could be profound. The warm water forming a part of the ocean conveyor belt of tropical Atlantic moves towards pole ward near the surface, when it gives up some of its heat to the atmosphere. This process partially moderates the cold temperature at higher latitude. The warm water gives up its heat and it become denser. This circulation loop is close as the cooled water makes its way slowly back towards the tropical at lower depths in the oceans. The melting of water from glaciers and the polar ice caps due to global warming can shut off ocean circulation and interrupt the ocean circulation system.

For a long time we did not take the oceans fully into account with regard to climate changes, because we had very little knowledge about the very complex phenomenon of oceans and ocean currents. In AD 1960 the approach regarding oceans and ocean currents began to change. The studies of clay of seabed indicated that the ocean current pattern might shift and other studies projected the complex and surprisingly fragile circulation of the global oceans. In the year 1980 the study of Greenland ice cores showed that the North Atlantic circulation could switch off radically within this century. The global warming might trigger such a switch which could wrek serious harms. The latest model of oceans also observed that this was unlikely to happen in this century.

The ocean circulation is comprised of global network of interconnected currents, counter currents, deep water currents and turbulent eddies. In this complex circulation an underlying transport pattern emerges. The water cycle from surface currents to deep water currents and then back to the surface again is called by the scientists the great conveyor belt or meridional overturning circulations (MOC). It is also known as the thermo-haline ocean circulations. Thermo means heat and haline means salinity. The reality is that without density driven process the deep water currents would no longer be created. The global great conveyor belt would grind to a halt. The sea water density depends on temperature and salinity. The salinity and temperature difference arise from heating/cooling at the sea surface and from the surface fresh water fluxes. The evaporation and sea ice formation enhance salinity; precipitation, run off and ice melt decrease salinity. The thermo-haline ocean circulation in contrast to the wind driven currents is not confined to surface water but can be regarded as a big overturning of the global oceans from top to bottom. The scientists are tracing the complex pathways of the MOC or great ocean conveyor belt. The oceans are rapidly warming and growing more acidic. The main ingredients of climate are not in the earth's atmosphere but lies in the oceans and ocean circulation/currents. Most of the water of Planet Earth lies in the oceans and even most of the gases dissolved in the ocean water. The thermo-haline circulation plays an important role in supplying heat to the polar regions and thus in regulating the amount of sea ice in these regions. The changes in the thermo-haline circulation are thought to have significant impacts on the earth's radiation budget. In so far as the thermo-haline circulation governs the rate at which deep waters are exposed to the surface, it may also play an important role in determining the concentration of carbon dioxide in the atmosphere.

The majority of scientists of the world believe that the temperature of Planet Earth is fastly rising due to the large scale production of anthropogenic green house gases. The temperature changes vary over the globe. The land temperature has increased since AD 1979 twice as fast as ocean temperature. The ocean temperature increase more slowly than land temperature because of the larger effective heat capacity of the oceans. The ocean lose more heat by evaporation. The northern hemisphere warms faster than the southern hemisphere because it has more landmass and more extensive areas of seasonal snow and sea ice cover subjected to the ice-albedo feedback. Although more greenhouse gases are emitted in the northern hemisphere than the southern hemisphere but this does not contribute in the difference in the global warming because the major green house gases persists long enough to mix between the hemispheres. The increase in carbon dioxide concentration from 280 ppm (parts per million) to 379 ppm from the year 1750 to 2007 is far greater than the natural increase, i.e. 180 ppm to 300 ppm over the last 650,000 years. This is the main cause of anthropogenic global warming. Some scientists believe that 2 degrees centigrade increase in the global mean temperature above pre-industrial level is a threshold beyond which the threat of major and irreversible damage becomes more plausible. The Planet Earth is heading towards catastrophic and irreversible crisis of climate change. The crisis is so fatal and serious that if it is not pushed back in short span of time, it would become uncontrollable and will push entire mankind along with other bio-life to their final doom, i.e. the extinction of the whole life on the Planet Earth. The mankind is heading towards a great global disaster. But till now we lack effective global disaster management in spite of great scientific-technological advances. The scientists, climate experts, environmentalists and rational social thinkers have given so many warnings to the governments of all the nations and world people about this dangerous environmental and human disaster. But the rich developed nations who are mainly responsible for the challenge of global warming/climate change and some big developing nations who are also emitting large scale green house gases are not ready to change their unsustainable corporate capitalist development model and to cut the green house gases at global level under global treaty because of their profit oriented sole aim.

Over the next 20 years, worries about climate change effect may be more significant than any physical changes linked to climate change. The rapidly changing Planet Earth may cause nations to take unilateral actions to secure resources, territory and other interests. Greater multilateral global cooperation is depend on number of factors such as behaviour of the countries, the economic context and the importance of the interests to be depended or won.

Many prominent international scientists worry that the recent assessments underestimate the impact of climate change and misjudge the likely time when effect will be felt. The scientists are still under process to understand the complex phenomenon of nature/environment. They have limited capability to predict the likelihood or magnitude of the extreme climate shifts but believed that it will not occur gradually or smoothly. The drastic cutbacks of GHG emissions probably would disadvantage the rapidly emerging economies. The global economy could be plunged into a great and very serious global recession or more worst global economic crisis. But if the present global corporate capitalist world will not cut effectively the GHG as it is required for the sustainability and the maintenance of proper balance of nature and mankind immediately, the whole human civilization and mankind including other bio-life on Planet Earth will face a great global disaster, which can eliminate the whole human race, human civilization and other bio-life on Planet Earth.

There is no scope of co-existence of unsustainable, anti-nature and anti-mankind global corporate capitalist system (which is creator of GHG and challenge of global warming/climate change and global dehumanization) and human race and other bio-life. Either global corporate capitalist system has to go or the entire human race and other bio-life to go from the scene of Planet Earth.

Time is very limited at our disposal and the whole mankind and all the governments of the world should immediately realize this truth and reality and firmly take the global decision without any confusion to replace the present anti-nature anti-mankind global corporate capitalist system through peaceful means and global mass movement. In order to save the beautiful human civilization, mankind, bio-life and build safer future for our generation, the establishment of nature-mankind centric/friendly global confederal new system is the demand of time.

It was only more than half century back the prominent scientists began to feel that something wrong is happening with environment and climate system of Planet Earth. The scientists and environment experts influenced the social thinkers and far sighted people about the imbalance of nature and mankind and the threat of global warming/climate change. They started to raise their voice that mankind should adopt sustainable social model (nature-mankind friendly model) sustainable development model and sustainable way of life. This is a time of great danger for our world and human civilization. Nothing short of fundamental and revolutionary global systemic change can save mankind and other bio-life on Planet Earth. Those who fail to grasp the new world reality and challenge of global climate change and global dehumanization will be pushed aside in the dustbin of history.

There are several reports issued by prominent scientists about global climate change:

1. In the year 2000, the statement issued by the 6000 internationally reputed scientists have emphasized that if the mankind did not stop the production of green house gases in a short span of time the global warming would make the Planet Earth uninhabitable for mankind and other bio-life by the end of 21st century.

2. In the year 2002, a study conducted by the UN World Wildlife Fund (WWF) has warned that if the mankind especially the rich nations and rich people do not change their extravagant and wasteful lifestyle, our Planet Earth will become unsustainable by 2050.

3. In the year 2003, Rome conference of the noted scientists of the world cautioned that if the worsening process of environmental degradation is not reversed within a short span of time, the Planet Earth will become totally unsuitable for mankind and other bio-life in the near future.

4. In the year 2004 a report of Natural Environment Research Council UK for ecology and hydrology stated that "six massive extinction events in the history of life are upon us yet again". Such mass extinction of living species of flora and fauna has periodically occurred during the five gigantic pre-historic extinctions. The UK centre said that previous five extinctions occurred due to the interplay of natural causes, while the sixth extinction seems to be due to human activity.

5. In the year 2004 UN report said that the smokeless industry of information technology is one of the major global pollutants. A new research has claimed that carbon dioxide level is rising at faster rate than the worst case. The scenario envisaged by the UN experts with the Planet Earth heading for the catastrophic and irreversible climate change by 2040.

Japan has struck by the most powerful earthquake-Tsunami on 11th March 2011. The 8.9 magnitude temblor, which was centred near the east coast of Japan which killed many thousands of people, caused the formation of 30 feet wall of water that swept across the rice fields, engulfed entire towns, dragged houses and buildings on to highways and tossed cars and boats like toys. Some waves reached 10 km in land in Miyagi prefecture on Japan's east coast. It was terrifying. The buildings shook, heaped and collapsed and numerous fires ignited. Thousands of people are missing the Japanese media reported. The PM of Japan Mr Naoto Kan said that the enormously powerful earthquake-tsunami had caused tremendous damage over a wide area.

The US geological survey said that the quake which struck at 2.46 pm on 11 March 2011 prompted the US National weather service to issue tsunami warnings for at least 50 countries and territories. The epicentre of main quake was located at Miyagi prefecture about 370 km north east of Tokyo. In Miyagi a train had derailed and authorities had lost contact with four trains in coastal areas. Six million households deprived from electricity. The Japanese police estimated that the death toll in quake-tsunami hit Miyagi prefecture could surpass 10000. The devastation stretched hundreds of miles along the coast, where thousands of hungry survivors huddled in darkened emergency centres, which are cut off from rescuers and aid. One report reveals that four whole trains along the coast had disappeared. The continued aftershocks hampered search efforts as the strong waves batter the coast line. The most worrying development was the danger of meltdown of nuclear reactor near the quake's epicentre. The nuclear reactors exploded and radiation started leaking. The explosion in nuclear reactors was caused by hydrogen interacting with oxygen outside the reactor. The hydrogen was formed when the super-heated fuel rods came in contact with water being poured over it to prevent a meltdown. Mark Hibbs a senior associate at the nuclear policy programme for the Carnegie Endowment for International Peace said that the Japanese are working furiously to find solution to cool the core and they have begun to inject sea water into the core. This is an indication of how serious the problem is and how the Japanese had to resort to unusual and improvised solution to cool the reactor core. The trouble began at the plant's unit no 1 after the massive 8.9 magnitude earthquake and the powerful tsunami which damaged its cooling system. The concerns about radiation leak of the nuclear power plant overshadowed the massive tragedy laid out along 1300 miles stretch of the coastline where scores of villages, towns and cities were battered by the tsunami packing 23 feet high waves. The tsunami was unbelievable fast. The US President Obama called the Japan's disaster a potentially catastrophic disaster.

In Japan 24% electricity is produced by 55 nuclear power units in 17 plants and some are in trouble after earthquake-tsunami. Japan declared state of emergency at power plants after their units lost cooling ability. The Japanese nuclear agency spokesman Shinji Kinjo acknowledged that there are still fears of a meltdown. A meltdown is not a technical term but rather it is an informal way of referring to a very serious collapse of a power plant's system and its ability to manage temperatures. Out of radiation fear everyone wants to get out of the town but due to devastation the roads are terrible. It was too dangerous to go anywhere because the wind may change direction and bring radiation anywhere.

Japan lie on the ring of fire – an arc of earthquake and volcanic zones stretching around the Pacific where about 90% of the world's quakes occur. In December 2004 quake occurred in Indian ocean that followed to dangerous tsunami killed an estimated 2,30,000 people in 12 countries. In February 2010 in central Chile an earthquake of 8.8 magnitudes occurred and generated tsunami killing 524 people. Technologically advanced Japan was well prepared for quake and its buildings could withstand strong jolts, even a temblor like 11 March 2011 which was the strongest the Japan ever experienced since official records started in the late 1800 century. What was beyond human control was the killer tsunami. The Japan's meteorological agency said on 13 March 2011 that it had upgraded the magnitude to 9.0. The agency earlier measured it at 8.8. The US geological survey has measured the quake at magnitude of 8.9. The Japan's worst previous earthquake was of a magnitude of 8.3 temblor in Kanto that killed 1, 43,000 people in 1923. The Japanese PM and emperor addressed the nation and asked the people not to lose hope. Both termed the earthquake-tsunami disaster as the worst crisis since World War Second. The International Atomic Energy Association is helping Japan in damage control and assessment of the nuclear hazards. The Fukushima nuclear power plant experienced explosion and with this there is a steep rise in the radiation level around it. The nuclear reactors exploded, leaking and causing harmful effects to health and dangerous exposure levels. The survivors of earthquake-tsunami in Japan are facing the prospect of a radiation leak damaging their health. The dangerous levels of radiation have been recorded near the Fukushima power plant and the radiation levels are rising in Tokyo. The nuclear radiation smashes up the DNA of cell. It can have both short-term and long-term effects on health. At very high levels it can cause radiation sickness. The radiation destroys the immune system. This leads to death within a few weeks. The lower doses can also be deadly in the long-term. Even small changes to DNA can lead to irreparable cell damage, causing cancer and birth defects. The long-term effects is an issue of serious concern. The radiation around the Chernobyl nuclear plant in 1986 caused a jump in thyroid cancer and leukaemia. Nearby Belarus showed 100 fold increase in thyroid cancer. In this situation what is to be done. The people should stay away from the affected areas and those in danger should take iodine tablets. The potassium iodide tablets provide protection against radiation. Staying indoors and wearing face masks offer only limited protection from larger particles of radioactive dust. The radioactive material is carried by minute moisture droplets in the air.

In Japan about 2,00,000 people have been evacuated from nearby sites. The radiation threat is becoming a scare among the people after the third explosion at the Fukushima nuclear plant. The nuclear reactor blast is turning out to be worse by the minute. The three radiation tests done near the reactor have been positive, which is alarming for the masses. The Japanese government admitted that the radiation is at that level which is very harmful to human health. In Japan nuclear emergency has been declared. The real assessment of the quake losses is yet to come but the latest estimates have surpassed everything else. After the biggest earthquake immediately a 30 meter tsunami wave hit the coast of Japan causing severe destruction. The tsunami has swept away everything that lay in its path-houses, buildings, aeroplanes, rails , bridges, streets, vehicles, farms and everything now lies undistinguishable from rubble.

The heat waves in Europe which killed 22,000 people, the cyclones Rita and Katrina in US which brought huge destruction and killings, earthquake-tsunami disaster in Indian ocean occurred in Indonesia 2004, unprecedented and disastrous floods in Pakistan and Australia and now disastrous earthquake-tsunami in Japan which brought huge destruction, great loss of human lives and ruined the economy of economic super power of the world. These natural disasters are clearly indicating the fury of nature, which is the outcome of natural causes as well as human induced causes. The global corporate capitalist system unscientifically and ruthlessly exploited and overused the natural resources and human resources for the purpose of maximum growth and maximum profit. The degradation of nature and dehumanization of mankind by the global corporate capitalist system created dangerous imbalance between nature and mankind and within mankind, resulting into the threat of global warming/climate change and global dehumanization. This threat is challenging the very existence of human civilization and mankind including other bio life. The nature is giving clear cut warnings through abov-said disaster to the whole mankind and the governments of the different countries of the world that don't interfere and violate the laws of nature/environment and stop the emission of greenhouse gases. The experience of these disasters and the study of degraded nature and dehumanized mankind guide us that the mankind should develop the human society with the tune of nature-mankind friendly global model and firmly build harmonious relations between nature and mankind and within mankind. The Planet Earth already lost the capacity to bear the burden of global corporate capitalist system due to overuse and over exploitation. The continue of anti-nature-anti-mankind global corporate capitalist system means further unbalancing the nature and mankind and to give definite invitation to global disaster, which will eliminate the whole human civilization and life on Planet Earth.

The global attention to climate change leads to unexpected impacts, thrusting the world into the new level of vulnerability. The corporate capitalist world adopted the growth first mentality leading to widespread environmental neglect and degradation. The corporate capitalist governments lost legitimacy as they have failed to cope up with the highly important environmental dimension. Despite the significant scientific-technological progress no technological silver bullet has been found to half the effects of climate change. The national solutions for global warming/climate change are short term and inadequate. The challenge of global climate change needs global solutions being a global problem.

The general public in several countries appear to be ahead of leaders in understanding the urgency of climate change threat or at least they had a better sense of the need for trade-offs. They have become early adopters for energy generation for renewable, the use of clean water technology and using improved internet connectivity to avoid the concentration of the people that make them vulnerable to extreme weather events. The Europeans have been out in the lead on energy efficiency but they have been too ready to sacrifice growth. In China it is opposite-too much crony capitalism and the public of China some years back was so grateful for the benefits according to China's hell bent efforts to modernize the Chinese society. But now it is different, the middle class wants clean air and water. They don't like the environmental devastation that was the price of rapid modernization.

Day-by-day the impact and the horrifying picture of deadly challenge of global warming and environmental degradation have become crystal clear. Then why the governments of different countries particularly the governments of most developed rich countries, which are 80% producers of green house gases and pollution, are indifferent towards the most dangerous fundamental challenge of global climate change? In fact, they are more concerned to attain maximum unsustainable growth, maximum profit and wealth instead of saving nature and mankind (the fundamental factors of human society) from the global disaster created by global warming-climate change. The first priority of corporate capitalist companies and the global corporate capitalist system is to achieve maximum growth and maximum profit without caring the sustainability of nature and mankind. The ongoing global economic recession (started from US, in 2007) has further hardened the global corporate capitalist unsustainable developmental standpoint and growth cum profit maximization priority. The global corporate capitalist system and corporate capitalist class are so callous and non serious that they are ready to loose mankind and human civilization but not their profits and crude economic interests. This is an inhuman approach, which is endangering the survival of human civilization and life on Planet Earth. There is a global need to conclude a legally binding rational global treaty, under which both developed and developing countries should reduce their emission of green house gases at that level which is necessary for the sustainability and survival of nature and mankind of Planet Earth. The rationality demands that the financial responsibility to tackle the crisis of global warming/

climate change and global dehumanization/human inequality should be shared mainly by the developed rich parts of the world and the rest of the world should also contribute as per their emission level and their financial capacity.

GLOBAL DEFORESTATION

The forest cover and bio-diversity are the important factors of environment of Planet Earth. The forests are the biggest carbon sinks, source of fresh oxygen after oceans and important supporting factor of rain. The forests and bio-diversity balance the natural ecosystem of Planet Earth and are the important elements of sustainability and survival of mankind. The anti-nature approach, human greed, over population, slash and burn farming, commercial agriculture, logging, growing lopsided urbanization, cutting of trees for fuel wood and one-sided unsustainable industrialization are the major responsible factors for global deforestation, loss of bio-diversity and aridity of Planet Earth. The original forest cover of Planet Earth was 15 to 16 million sq km. But now 50% (8 million sq km) mature tropical forests had been destroyed. The rain forests 50 years back use to cover 14% land surface but now they have reduced up to 6%. The earth's forest cover since 1990 has shrunk by more than 7 million hectares each year, with annual losses of 13 million hectares in developing countries and re-growth of almost 6 million hectares in industrial countries. To protect the Planet Earth's nearly 4 billion hectares of remaining forests and replanting those already lost, are essential for restoring the earth's health. There is an estimate that if the present trend of deforestation will continue then all the tropical forests of Planet Earth will be destroyed by 2090. The developed countries have destroyed their most of the forest cover many years back and they attained huge economic benefit out of deforestation. The developing countries also destroyed their forest cover in large extent and they are still damaging their forests and bio-diversity. The most of the countries of the Planet Earth have destroyed large areas of their rain forests and tropical forests. The continuous destruction of forests and bio-diversity has created great ecological and environmental imbalance. There is an estimate that 30% of natural species will extinct by 2050. The degradation of forest cover (both tropical forests and rain forests) of Planet Earth is posing a great environmental problem. It is also contributing to global warming/climate change. The global deforestation damage bio-diversity, affects the climatic conditions and creates deserts, do adverse impact on bio-sequestration of atmospheric carbon dioxide and displacement of indigenous people. The reducing rainfall runoff and the associated soil erosion and flooding, and restoring aquifers recharge depends on both forest protection and reforestation. The reforestation can reduce pressures on the earth's remaining forests. The South Korea is a model of reforestation for the rest of the world.

ROLE OF UN

The UN has taken several international initiatives on global challenge and threat of global warming/climate change. The number of high level international conferences has been held under UN since the last 40 years on sustainable development and global warming/climate change in order to chalk out a workable strategy to counter the impending catastrophe but all these international conferences/summits could not achieve the targets and goals. The major international conferences/ summits on sustainable development and environment held under UN are:

1. The Stockholm Conference on sustainable development and environment, 1972 (Sweden)
2. Helsinki Conference, 1989 (Finland)
3. London Conference, 1992 (UK)
4. Rio-de Jinero Earth Summit, 1992 (Brazil)
5. Kyoto Summit, 1997 (Japan)
6. Johannesburg world Summit on sustainable development, 2002 (South Africa)
7. Stockholm Summit, 2004 (Sweden)
8. Montreal Summit, 2005 (Canada)

9. UN World Summit on global warming and climate change at Bali, 2007 (Indonesia)
10. The Copenhagen Global Conference on climate change, December 2009 (Denmark)
11. Cancun Global Summit on Climate Change, 2010 and
12. Durbin Global Climate Conference 2011.

But all these UN global conferences could not attain the desirable results.

The role of UN about global warming/climate change and sustainable development from the past four decades are as under: The United Nations Framework Conventions on Climate Change (UNFCCC) was opened for signature as the 1992 UN Conference on Environment and Development (UNCED) in Rio de Jinero (known as earth summit) on June 12, 1992, in which 154 nations signed the UNFCCC in order to reduce atmospheric concentration of green house gases with the goal of preventing the dangerous anthropogenic interference with the earth's climate system. These actions were aimed primarily at industrialized countries, with the intention of stabilizing their emissions of green house gases at 1990 levels by the year 2000 and other responsibilities would be incumbent upon all the UNFCCC parties. The UNFCCC parties agreed in general that they would recognize common but differentiated responsibilities, with greater responsibilities to reduce green house gas emission in the near term on the part of developed/industrialised countries which were listed in Annexure 1 of the UNFCCC. The UNFCCC have received over 50 countries' instruments of ratification, it entered into force on March 21, 1994. Since the UNFCCC entered into force, the countries have been meeting annually to Conferences of the Parties (COP) to assess progress in dealing with climate change and beginning in the mid-1990s, to negotiate the Kyoto Protocol to establish legally binding obligations for developed industrialized countries to reduce their green house gases emission. The Kyoto Protocol (1997) which was signed by 140 nations and almost majority of nations has ratified but the US in spite of the signatory, did not ratify the Kyoto Protocol and weakened this global agreement which was legally binding agreement on most industrialized rich corporate capitalist nations to reduce their green house gases emission up to the 1990 level.

UNIPCC SUMMARY REPORT

The global climate change is the defining human development challenge of the 21st century. The failure to respond global climate challenge will endanger human community and whole bio-life on our Planet Earth. The poorest countries and the most vulnerable citizens will suffer the earlier and most damaging setbacks, even though they have contributed least to the problem. Looking to the future no country—however wealthy or powerful—will be immune to the impact of global warming/climate change. The UN IPCC's summary report shows that the climate change is not just a future scenario. The increased exposure to the droughts, floods and storms are already destroying opportunity and reinforcing danger. Meanwhile, there is now overwhelming scientific evidence that the catastrophe become unavoidable. The climate change point out in a clear direction: unprecedented reversal in human development in our lifetime and acute risks for our children and their grandchildren.

There is a window of opportunity for avoiding the most damaging climate impact, but that window is closing – the world has less than a decade to change course. Action taken or not taken in the year ahead will have a profound bearing on the future course of the human development. The world lacks neither the financial resources nor the technological capability to act. Missing is human solidarity and a sense of collective interest.

The UNIPCC report argues the climate change poses challenges at many levels. In a divided but ecologically inter-dependent world, it challenges all people to reflect upon how we manage the environment of one thing that we share in common Planet Earth. It challenges political leaders and people in rich nations to acknowledge their historic responsibility for the problem and to initiate deep and early cuts in green house emissions. Above all it challenges the entire human community to undertake prompt and strong collective action based on shared values and shared visions. The global conferences [Conferences of Parties] on global climate change held under UNFCCC are as under:

COP 1 The Berlin Mandate Germany (1995)

The first UNFCCC Conference of Parties took place in the year 1995 in Berlin, Germany. It voiced concerns about the adequacy of countries' abilities to meet commitments under the Convention. In Berlin conference a ministerial declaration was signed [which is known as Berlin Mandate] to negotiate a comprehensive menu of actions for countries to choose future options to address the challenge of climate change. The Berlin Mandate exempted non-annex I countries from additional binding obligations, in keeping with the principle of "common but differentiated responsibilities" established in the UNFCCC.

COP 2 Geneva Switzerland (1996)

The COP 2 took place in July 1996 in Geneva [Switzerland]. In the Geneva conference the ministerial declaration was adopted on July 1996 and reflected the US position statement as under: (1) Accepted the scientific findings on climate change proffered by the Intergovernmental Panel on Climate Change (IPCC) in its second assessment (1995); (2) Rejected uniform "harmonized policies" in favour of flexibility; (3) Called for "legally binding mid-term targets."

COP 3 The Kyoto Protocol, Japan (1997)

The COP 3 took place in December 1997 in Kyoto, Japan. After intensive negotiations, it adopted the Kyoto Protocol. The most industrialized nations and some central European countries under Kyoto Protocol agreed for legally binding reductions in greenhouse gas emissions of an average of 6 to 8% below 1990 levels between the years 2008 and 2012. The United States would be required to reduce its total emissions on an average of 7% below 1990 levels, however neither the Clinton administration nor the Bush administration sent the Kyoto Protocol to US Congress for ratification. The Bush administration explicitly rejected the Kyoto Protocol in 2001. The Barrack Obama also refused to follow the Kyoto Protocol and sponsored new climate order in Copenhagen Climate Conference with four developing countries (China, India, Brazil, and South Africa).

COP 4 Buenos Aires, Argentina (1998)

The COP 4 took place in November 1998 in Buenos Aires. It had been expected that the remaining issues unresolved in Kyoto would be finalized at this conference. However, the complexity and difficulty of finding agreement on these issues proved insurmountable and instead the parties adopted a two-year "plan of action" to advance efforts and to devise mechanisms for implementing the Kyoto Protocol by year 2000.

COP 5 Bonn Germany (1999)

The COP 5 took place between October 25 and November 5, 1999, in Bonn, Germany. It was primarily a technical meeting and did not reach any major conclusion.

COP 6 The Hague Netherlands (2000)

The COP 6 took place between November 13–25, 2000, in The Hague, Netherlands. In the Hague conference the discussions evolved rapidly into a high-level negotiation over the major political issues. These included major controversy over the United States' proposal to allow credit for carbon "sinks" in forests and agricultural lands, satisfying a major proportion of the US emission reductions in this way; disagreements over consequences for non-compliance by countries that did not meet their emission reduction targets and difficulties in resolving how developing countries could obtain financial assistance to deal with adverse effects of climate change and meet their obligations to plan for measuring and possibly reducing greenhouse gas emissions. In the final hours of COP 6, despite some compromises agreed between the United States and some EU countries, notably the United Kingdom, the EU countries as a whole, led by Denmark and Germany, rejected the compromise positions and the talks in The Hague collapsed. Jan Pronk, the President of COP 6, suspended COP-6 without agreement, with the expectation that negotiations would later resume. It was later announced that the COP 6 meetings (termed "COP 6 bis") would be resumed in Bonn, Germany, in the second half of July. The next regularly scheduled meeting of the parties to the UNFCCC - COP 7 — had been set for Marrakech, Morocco, in October-November 2001.

COP 6 Bis, Bonn Germany (2001)

The COP 6 negotiations resumed in July 17–27, 2001, in Bonn, Germany, with little progress having been made on resolving the differences that had produced an impasse in The Hague. However, this meeting took place after President George W. Bush had become the US President and he rejected the Kyoto Protocol in March; as a result the United States delegation to this meeting declined to participate in the negotiations related to the protocol and chose to act as observers at that meeting. As the other parties negotiated the key issues and agreement was reached on most of the major political issues, to the surprise of most observers given the low level of expectations that preceded the meeting. The agreements included:

1. *Flexible mechanisms:* The "flexible" mechanisms which the United States had strongly favoured as the protocol was initially put together, including emissions trading, joint implementation and the clean development mechanism (CDM) which allow industrialized countries to fund emissions reduction activities in developing countries as an alternative to domestic emission reductions. One of the key elements of this agreement was that there would be no quantitative limit on the credit.

2. *Sinks:* The credit was agreed for broad activities that absorb carbon from the atmosphere or store it, including forest and cropland management and re-vegetation, with no overall cap on the amount of credit that a country could claim for sinks activities.

3. *Compliance:* The final action on compliance procedures and mechanisms that would address non-compliance with protocol provisions was deferred to COP 7, but included broad outlines of consequences for failing to meet emissions targets that would include a requirement to "make up" shortfalls at 1.3 tons to 1, suspension of the right to sell credits for surplus emissions reductions and a required compliance action plan for those not meeting their targets.

4. *Financing:* Three new funds were agreed upon to provide assistance for needs associated with climate change; a fund for climate change that supports a series of climate measures; the least developed country fund to support national

adaptation programs of action and Kyoto Protocol adaptation fund supported by CDM levy and voluntary contributions. The number of operational details attendant upon these decisions remained to be negotiated and agreed upon and these were the major issues of the COP 7 meeting that followed.

COP 7 Marrakech Morocco (2001)

The COP 7 was held at Marrakech from October 29 to November 10, 2001. The negotiators completed the work of the Buenos Aires plan of action. The COP7 finalized most of the operational details and set the stage for nations to ratify the protocol. The complete package of decisions is known as the Marrakech Accords. The United States delegation continued to act as an observer, declining to participate in active negotiations. The other parties continued to express their hope that the United States would re-engage in the process at some point, but indicated their intention to seek ratification of the requisite number of countries to bring the protocol into force. A target date for bringing the protocol into force was put forward: the August-September 2002 world summit on sustainable development (WSSD) to be held in Johannesburg, South Africa. The main decisions at COP 7 included:

1. Operational rules for international emissions trading among parties to the protocol and for the CDM and joint implementation.
2. The compliance regime that outlines consequences for failure to meet emissions targets but defers to the parties to the protocol after it is in force to decide whether these consequences are legally binding.
3. Accounting procedures for the flexibility mechanisms.
4. The decision to consider at COP 8 that how to achieve a review of the adequacy of commitments that might move toward discussions of future developing countries commitments.

COP 8 New Delhi India (2002)

The COP-8 under UN Framework Convention on Climate Change met in New Delhi from October 23 to November 1,

2002, in conjunction with the seventeenth sessions of the subsidiary body on scientific and technological advice and the subsidiary body on implementation. The most of the issues relating to implementation rules for the Kyoto Protocol resolved at COP-7 in Marrakech but the protocol was not yet in force, the formal agenda at COP-8 comprised mostly of second-order and technical issues. However, beyond the formal agenda in political statements and in halfway discussions the COP-8 also saw the emergence of a vigorous debate over next steps in the development of the climate change regime. The wide differences among parties on that question was reflected in the negotiations over the Delhi declaration, a broad political statement meant to reflect the consensus among parties at COP-8. The United States, while reiterating its opposition to the Kyoto Protocol, was deeply engaged in the negotiations as a party to the framework convention and as a member of the umbrella group (developed countries outside the European Union and Eastern Europe). The lack of developing countries commitments as a primary basis for its rejection of Kyoto Protocol, the United States struck a far different tone in Delhi, declaring that it would be "unfair" to insist that developing countries adopt greenhouse gas targets. The United States also pressed hard on a number of issues that appeared to take on broader significance as a test of other parties' willingness to accommodate US concerns. The emphasis in COP 8 for many parties was the importance of bringing Kyoto Protocol into force as quickly as possible.

Many of the issues were deferred for further consideration at future meetings. The outcome of Cop 8 are as under:

1. Adopted the Delhi ministerial declaration on climate change and sustainable development.
2. Adopted rules of procedure for the executive board of the clean development mechanism (CDM).
3. Completed work on the reporting required of developed countries to assess their compliance under the Kyoto Protocol.
4. Adopted guidance to the global environment facility (GEF) for managing two new funds established at COP-7 to assist developing countries.

5. Adopted new guidelines for national communications to be submitted by developing countries reporting on their emissions and steps they are taking to meet their commitments under the framework convention.
6. Requested the Intergovernmental Panel on Climate Change (IPCC) and the Montreal protocol's technological and economic assessment panel to conduct a special report on the question of HFCs/PFCs—compounds that have replaced ozone-depleting substances but contribute to climate change (details of Delhi declaration can be read at annexure-I).

COP 9 Milan Italy (2003)

The COP 9 [Conference of Parties] was held in Milan, Italy, in December 2003. The Milan Conference produced modest progress on a handful of largely technical issues but remained essentially deadlocked on the issues touching on the broader question of next major steps in the international climate effort. The talks formally known as ninth session of the conference of the Parties under UN Framework Convention on Climate Change came against the backdrop of continued uncertainty over the fate of the Kyoto Protocol. During the first week of the COP 9, there emerged from Moscow another round of conflicting signals on the prospects for Russian ratification of the protocol and thus its entry into force. In Milan, nevertheless, most of the parties reaffirmed their strong support for Kyoto Protocol and remained publicly hopeful that Russia will ratify. The important outcome of Milan Conference was, the decisions on the technical rules for sink projects in the clean development mechanism and on guidelines for the operation of two funds to assist developing countries: the special climate change fund and the least developed countries fund. The Milan Conference was attended by 100 ministers and they participated in three loosely framed roundtable discussions that served largely as an opportunity to restate familiar positions. The result in most cases was continued stalemate, with decisions only to discuss the issues further in future negotiations. In both public statements and private remarks, delegates expressed a mix of deepening frustration with the negotiating process and new openness to alternative approaches both within and outside the UNFCCC framework.

COP 10 Buneos Aires Argentina (2004)

The COP 10 held at Buneos Aires, Argentina, on 6–17 December, 2004. The COP 10 convenes as the international climate change effort enters a critical transitional phase. In November 2004, the Russia ratified the Kyoto Protocol. The final step needed to bring the Treaty into force. The protocol establishes binding greenhouse gas emission limits for the 30 industrialized countries that have ratified it and an international emissions trading system. The Kyoto Protocol's entry into force also sets the stage for negotiations beginning in 2005 towards future climate commitments. The COP 10 serves largely as an opportunity to assess progress and consider the challenges ahead. The broad theme of the high-level segment attended by ministers was the 10th anniversary of the UNFCCC's entry into force. The one possible outcome was agreement to organize a seminar in 2005 to begin exploring possible next steps in the international efforts. The other issues include efforts by developing countries to accelerate assistance for adaptation to climate change, the treatment of sinks in the protocol's clean development mechanism and the treatment of emissions from marine and aviation fuels.

COP 11 Montreal Canada (2005)

The COP 11 took place between November 28 and December 9, 2005, in Montreal, Quebec, Canada. The COP 11 was also the first meeting of the parties (MOP-1) to the Kyoto Protocol since their initial meeting in Kyoto in 1997. It was, therefore, one of the largest intergovernmental conferences on climate change ever. The event marked the entry into force of the Kyoto Protocol and hosting more than 10,000 delegates. It was one of the Canada's largest international events ever and the largest gathering in Montreal. The Montreal action plan is an agreement hammered out at the end of the conference to "extend the life of the Kyoto Protocol beyond its 2012 expiration date and negotiate deeper cuts in greenhouse-gas emissions." The Canada's environment minister, at the time, Stéphane dion, said the agreement provides a "map for the future."

COP 12 Nairobi, Kenya (2006)

The COP 12 took place between 6 and 17 November 2006 in Nairobi, Kenya. In the conference, the phrase climate tourists were coined to describe some delegates who attended to see Africa, take snaps of the wildlife, the poor, dying African children and women.

COP 13 Bali, Indonesia (2007)

The COP 13 took place on December 3 to 15, 2007, at Nusa Dua, in Bali, Indonesia. The agreement on a timeline and structured negotiation on the post-2012 framework (a successor to the Kyoto Protocol) was achieved with the adoption of Bali Action Plan. The ad hoc working group on long-term cooperative action under the convention was established as a new subsidiary body to conduct the negotiations aimed at urgently enhancing the implementation of the convention. In Bali (Indonesia) global climate conference resolved to urgently enhance the implementation of the convention, in order to achieve ultimate objective and goal in accordance with its principles and commitments. It was also decided to address the global challenge of global warming/climate change as it was reflected in the fourth assessment report of the intergovernmental panel on climate change. In the Bali global climate conference it was pledged and formulated in the Bali Action Plan to launch a comprehensive process to enable effective, sustained and full implementation of the convention (UNFCCC) through long-term cooperative action, up to and beyond 2012, in order to reach an agreed outcome. The common vision for cooperative action and long-term global goal of green house gases emission reduction and to achieve the ultimate objective of the convention with common but differentiated responsibilities and respective capabilities, keeping in view the social- economic conditions and other relevant factors. In Bali Action Plan it was also decided to enhance national and international action on mitigation and adaptation of climate change. To enhance action on technology development and technology transfer, in order to support action on mitigation and adaptation.

COP 14 Poznan, Poland (2008)

The 2008 United Nations Climate Change conference COP 14 in Poznan Poland, took place from 1 Dec to 12 Dec, 2008. The delegates agreed on principles of financing for a fund, to help the poorest nations cope with the effects of climate change and also they approved a mechanism to incorporate forest protection into efforts.

COP 15 Copenhagen, Denmark (2009)

The Copenhagen (Denmark) global climate conference (COP 15) held under UNFCCC on 7th to 18th December, 2009. The number of heads of the governments, heads of the states, ministers, delegates of 193 countries, the delegates of different environmental organizations and delegates of NGOs participated in the Copenhagen global climate conference. The 12 day's contentious and tough negotiations on global warming/climate change could not reach at any consensus agreement. The whole corporate capitalist world was divided on the issue of green house gases cutbacks. The US unilaterally announced 17% cutbacks, China 40% to 45% up to 2020 (China's state council announced that it would reduce the amount of carbon dioxide it emits per unit of GDP by 40–45% below 2005 levels by 2020). France and Brazil adopted a common climate goal. Brazil offered to volunteer cut in green house gases emission between 36 and 39% before the Copenhagen climate conference. The India declared to cut the carbon emission intensity by 20–25%. The Commonwealth countries in their 2009 summit at Trinidad tried to chalk out common strategy on climate change before Copenhagen global conference on climate change. The US president Barack Obama prior to Copenhagen climate conference, tried to get support of China, India, South Africa, Brazil, etc. to favour US stand in Copenhagen global climate conference. The unilateral declaration of cutbacks of green house gases under corporate capitalist and narrow nationalistic approach by some countries and the US corporate capitalist cum superpower approach damaged the cause of Copenhagen global climate conference. The global warming/climate change is not a national phenomenon but a global threat to whole mankind and other bio-life. The inept and dubious negotiations that have led China, India, Brazil, and South Africa to endorse US sponsored Accord. China, India, Brazil and South Africa choose to identify themselves with US instead of, the majority of poor and underdeveloped nations. They forgot their previous commitments, declarations and statements in favour of Kyoto Protocol, Bali Action Plan and legally binding global treaty. These big five nations US, China, India, Brazil and South Africa disheartened the G77 nations in particular and global people in general. The African countries expressed their opinion that the developed and big developing countries are generating more than 80% green house gases and they should financially and technologically contribute to control global warming/ climate change. The majority of the countries and global people did not accept the US backed five nations Accord. They out rightly rejected this Accord. It was the gross violation of global spirit and democratic majority norm. The Copenhagen US backed Accord only passed by a procedural motion after two weeks of tense negotiations. The Accord has been widely condemned as a back door deal that excluded the poor countries and doomed the world to disastrous climate change. The agreement was assembled and reached among the leaders of US, China, India, Brazil, South Africa, and major European countries when it become clear that the climate summit was in danger of failure. But in fact, the Accord was openly declared by US, China, India, Brazil, and South Africa. In reality, the conference utterly failed and ended without legally binding global agreement (except unapproved US backed Accord) which is the historical need for the survival of mankind. The US succeeded to influence China, India, Brazil and South Africa in its favour to toe its line (which is contrary to the global expectations of global people) and divided them from poor and developing countries. China, India, Brazil and South Africa by supporting the last moment US sponsored Accord, disappointed and hurt the poor and developing countries and the global people. The Accord was not approved by the majority of the countries presented in the Copenhagen global climate conference. The Accord was merely informed to the 193 countries at the Copenhagen Climate Conference in the last days, rather than getting approval, which would have required unanimous support. It was an undemocratic and back door act. The Accord is extremely weak in terms of deep and immediate emission cuts needed to be taken by the developed countries. Though the fiasco in Copenhagen climate conference was averted with face saving Accord. The UN Secretary General Ban Ki Moon said it needs to be turned into a legally binding treaty next year (2010). He even admitted that the Copenhagen Climate Conference had failed to win global consensus and disappointed many who demanded stronger action against global warming/climate change. The chairman of UNIPCC Dr. Rajinder Pachouri said that the developing countries certainly the African countries are very suspicious about the developed countries, whether they are really genuine in making these offers. We will have to make sure that it will move quickly towards the status of a legally binding agreement. We really have to move on rather quickly to reduce the emission of green house gases. There is growing evidence of the impacts of climate change. If we delay action then these impacts are going to become much worse and far more serious. The Accord projected the need to limit global temperatures rising no more than 2 degrees Celsius above pre-industrial levels. The Accord promises to deliver 30 billion dollars of aid for developing poor countries over the next three years. It outlines a goal of 100 billion dollars by year 2020 to help poor countries to cope with the impacts of climate change. The Accord says the rich countries will jointly mobilise the US $100 billion and a green climate fund will also be established to support projects in developing countries related to mitigation, adaptation, capacity building and clean technology transfer. The pledges of rich countries will come under "rigorous, robust and transparent scrutiny under UNFCCC. Under the Accord the developing nations will submit national reports on their emission pledges under a method that will ensure that national sovereignty is respected. The implementation of the US backed Accord will be reviewed by 2015. This will take place about one and a half years after the next scientific assessment of the global climate by the UNIPCC".

The US president Barack Obama described this Accord as the meaningful agreement and would need to be built on in the future and that we have come a long way but we have much further to go. The UK PM Gordon Brown said, "We have made a start but that the agreement needed to become legally binding quickly". The European union president Jose Manuel Barroso said, "I will not hide my disappointment regarding the non-binding nature of the agreement". The French president Nicholas Sarkozy commented, "The Accord we have is not perfect". The China's PM wen Jiabao said that the weak agreement was because of distrust between nations and to meet the climate change challenge, the international community must strengthen confidence, build consensus, make vigorous efforts and enhance cooperation. The Brazil's climate change ambassador called the agreement disappointing. The head of the G77 group of countries said that the US backed Accord asked African countries to sign a suicide pact and that it would maintain the economic dominance of a few developed rich countries. The representatives of the Maldives, Venezuela and Tuvalu, etc. were unhappy with the outcome. The Venezuelan representative said that international agreement cannot be imposed by a small exclusive group. It is a coup d'état against UN. The Bolivian president Evo Morales said that the Copenhagen global climate conference has failed and it is unfortunate for the Planet Earth. The fault is with the lack of political will by a small group of countries led by the US. The John Ashe, the chairman of the talks that led to the Kyoto Protocol, was also disappointed with the agreement and said that the expectations for this global climate conference, anything less than a legally binding and UN backed agreed outcome falls far short of the mark.

The John Sauven executive director of green peace UK stated that the city of Copenhagen was a crime scene. It is now evident that beating global warming/climate change will require a radically different model of politics than the one on display here in Copenhagen. The Nimmo Bassey friends of the earth International call the Copenhagen Climate Conference an abject failure. The Lydia Baker of save the children said that world leaders had effectively signed a death warrant for many of the world's poorest children and 250,000 children from poor communities could die before the next climate conference in Mexico at the end of 2010. The Tim Jones climate policy officer from the world development movement said that leaders had refused to lead and instead sought to bribe and bully the developing nations to sign up to the equivalent of a death warrant. The Kim Carstensen leader of climate initiatives WWF for nature stated that the Accord is well meant but half-hearted pledges to protect Planet Earth from dangerous climate change. It is not sufficient to address the crisis that calls for completely new ways of collaboration across rich and poor countries. The Robert Bailey of Oxfam international said, "It is too late to save the Copenhagen Climate Summit but it is not too late to save the Planet Earth and mankind. We have no choice but to forge forward towards a legally binding global treaty in 2010. This must be a rapid, decisive and ambitious movement, not business as usual.

The Copenhagen global climate summit (Dec. 2009) despite wide expectations that the summit would produce a legally binding global treaty, was plagued negotiating dead lock and the US backed undemocratic Accord is not legally enforceable. The failure of the Copenhagen global climate conference attributed due to the number of factors, such as the recent global economic recession and conservative domestic pressure in US, China and other developed and some developing countries. The US backed Accord asks countries to submit emissions targets by the end of January 2010 and paves the way for further discussion to occur at the December 2010 UN Climate change conference in Mexico and the mid-year session in Bonn, Germany. The future of the UN role in global climate agreement is now in doubt.

The Accord negotiated by US president Obama with the leaders of China, India, Brazil, and South Africa did not meet even the modest expectations that set for Copenhagen climate conference (COP 15). The Accord does not firmly commit from the Industrialized nations or the developing nations to firm targets for mid-term or long-term green house gases emission reduction. It is an equivocal agreement, a disappointing conclusion. The messy compromise mirrored the chaotic nature of the Copenhagen global climate conference. The Accord would have been almost worthless without recognition by the plenary session of the entire delegate nations at the global climate summit. The democratic multilateral process of the UNFCCC has been replaced by the will of the powerful. The coup d'état that the Copenhagen Accord. The Latin American and African countries stood to the pressure and ensured an outcome whereby COP 15 was forced to merely take note of the Copenhagen Accord without adopting or endorsing it. The way forward remains mystery. The burial of the Rio Declaration (1992), the UN framework Convention on Climate Change (1992), Kyoto Protocol (1997) and the Bali Action Plan (2007) has begun. By accepting US sponsored Accord through US president Obama and abandoning commitments made under Bali Action Plan, the China, India, Brazil and South Africa have blessed a global warming of at least 3 degrees Celsius at little cost to the rich countries. The great historical responsibility and the touch stone of the UN framework convention do not find any reference in the US backed Accord of five. The Accord recognizes the 2 degrees Celsius goal but does not endorse it. The Accord opens up nationally appropriate actions in place of International consultation, analysis along with global action. There is no clarity on the way forward other than an assessment of the Accord's implementation by 2015. The entire global consensus built up since Rio Earth Summit (1992), Kyoto Protocol (1997), Bali Action Plan (2007) and post-Bali Conference exercises have been set aside to the dictate of new climate world order dominated by the US, developed and big developing countries. The US emissions have already usurped 30% of the available global environmental space to support consumption of less than 5% out of the world population. The US remained the world's largest emitter till 2007, when it was overtaken by China. If one accounts for the embedded emissions in the net imports guzzled by the US consumers, the US remains the largest emitter by far even today. The US and China account for almost half the world's emissions. They must do their part. If they don't fulfil their responsibility we will not be able to meet 2 degrees Celsius target. The China's share of available global environmental space is almost 2.5 times that of India. If the world leaders particularly the leaders of developed and big developing countries will not mend their negative approach and attitude early, then the time is coming when the global people will sharply react in the form of global democratic mass movement against govts and leaders of these countries and will unitedly raise their voice against global warming/climate change and global dehumanization and its creator the global corporate capitalist system. The George Monbiot blamed the failure of Copenhagen Climate Conference to achieve a legally binding global treaty on the US Senate and President Barack Obama. By negotiating the Copenhagen Accord by only a select group of nations, most of the UN member states were excluded.

India, China, Brazil and South Africa, all emerging economies co-operated US at Copenhagen to thwart attempts at establishing legally binding targets for carbon emissions in order to protect their economic growth. The US backed Accord reflects that the US, developed countries and four developing countries that supported this Accord are more concerned about their corporate capitalist economies rather than to save the nature and mankind from the danger of global warming/climate change. The US, China, India, Brazil and South Africa should rise above the national and corporate capitalist interests and review the Accord and firmly stand with poor developing countries along with global people for legally binding global climate treaty.

Now the world is mainly divided into most developed rich countries and poor developing countries (those are least responsible for the creation of the global warming/climate change) are demanding from the Industrialized developed countries to provide technology and funds to face the challenge of global warming/climate change. The UN adaptation fund under Kyoto Protocol collected 3.5 billion dollars for poor countries (in the year 2001) to tackle the green house gases emission. But up to 2009 not even a single poor country has got the funds for controlling the global warming/climate change under UN adaptation fund. The UN sponsored carbon credit emission policy to give economic benefit to those corporate companies of the world who will reduce their carbon emission, is also a corporate capitalist money prescription not the real solution of the problem of global warming/climate change. The UNFCCC in its meeting held on November 2007 issued strongest warning on climate change that the world governments only have few years left to avert the worst impact of the challenge of global warming/climate change.

Under this state of balance of global power what is to be done? It is only the united pressure of the global people which can force the global corporate capitalist governments to sign legally binding global climate treaty on global warming/ climate change and save the nature and mankind from this global disaster.

Tianjin Meeting on Climate Change

The working group meeting on global climate change under the United Nations Framework Convention of Climate Change and the Kyoto Protocol held on 6th to 9th October 2010 at Tianjin, China. In the meeting three thousand delegates participated from 176 countries in order to make progress on global climate agreement and set the tone for the annual UNFCCC climate summit which will be held from November 29th to 10th December in Cancun, Mexico. The Government negotiators and non-governmental organizations said that the Tianjin and Cancun meetings could produce enough progress for legal decisions that launch a global system to preserve forests, establish a financial system to coordinate adaptation and emissions reduction projects, formalize emissions reductions commitments and establish a system to review progress toward those goals. The challenge is whether the Cancun Climate Summit will set a stronger negotiating foundation, which can lead to a formal International agreement between all the nations in time for the 2011 climate summit in South Africa or the 2012 summit in South Korea. The United States and China, the two largest carbon polluters, have been reluctant to set a binding global emissions limit. There is evidence that the maximum nations want to reach that goal. Last year in Copenhagen, [Denmark] the heads of state and climate negotiators from 28 countries that are responsible for 80 percent of climate changing emissions developed the Copenhagen Accord. The Accord was a political agreement which indicates that the nations should individually and collectively check carbon emissions sufficiently to limit the rise in global temperature to less than 2 degrees centigrade. In the Accord, which was reached on the last day of the Copenhagen Summit, the developed nations also committed to provide US $30 billion from 2010 to 2012 to assist developing nations to make the transition from a carbon-based to a clean energy economy.

The UNFCCC reported that 139 nations, including the 27 member European Union, have agreed to the Accord or have expressed their intent to sign on; The UNFCCC reported in March that it received submissions of national pledges to cut or limit emissions of greenhouse gases by 2020 from 75 countries, which account for all but 20 percent of global emissions. In addition 41 industrialized countries formally communicated their economy-wide targets and 35 developing countries have "communicated information on the nationally appropriate actions they are planning, provided they receive the appropriate support in terms of finance and technology. The China and the United States sucked up most of attention and concern. India is acting to reduce its carbon emissions and expanding solar power, Mexico is adopting new vehicle efficiency standards and the European Union invested US $41 billion last year in clean energy and more than half of the continents new power production in 2009 came from renewable energy sources. But the China and US still the centre of attention who are pursuing a kind of energy and climate strategy that by necessity commands the world's focus. China's energy consumption last year was the world's highest, equivalent to 2.3 billion tons of oil, 0.4 percent more than the US energy consumption, according to the international energy agency. Moreover, demand for energy is rising faster in China than any other country and it is the largest producer of climate emissions, according to the energy information administration, a unit of the US department of energy. The Chinese officials said that they are meeting the goal, announced last year in Copenhagen, of a 40 to 45 percent reduction in carbon intensity by 2020, an analysis by the natural resources defence council concludes that China's carbon emissions will essentially double over the decade to more than 12 billion tons a year. The basic reason is that even as China invested US $35.6 billion last year in clean energy development, announced closures of hundreds of inefficient coal-fired power plants and became the unmistakable global leader in solar manufacturing, it also said it will increase coal production by 2020 to over 4.2 billion metric tons annually, an increase of 33 percent from the 3.15 billion tons, the country will mine and consume this year, according to the China national coal association. The China produces and burns more coal than any other country-three times more than the US and coal supplies 70 percent of the nation's total energy demand. The US meanwhile passed legislation in 2009 to invest roughly US $100 billion in clean energy development, energy efficiency, and fuel saving transit and became the largest generator of wind energy in the world. The American oil industry is spending an estimated US $100 billion annually to perpetuate the fossil fuel era by developing "unconventional" sources of oil and natural gas from tar sands and deep shales that produce more carbon emissions than "conventional" fuel sources. In US and China the negotiations about limiting carbon emissions is impeded by economic priorities. Clearly both the countries embrace the existing fossil fuel economy as essential to national stability and well-being. Both countries also are pursuing clean energy development; China produces seven percent of its energy from renewable sources; the US produces eight percent.

The Tianjin conference tried to expand common ground and lessen disagreement to lay out the negotiation text for the Cancun meeting.

COP-16 Outcome of Global Climate Conference at Cancun Mexico

The 16th UN global climate conference under UNFCCC held at Cancun Mexico on 29th Nov to 10 Dec 2010. In this global climate conference the representatives from 194 nations, the representatives of various NGOs and prominent international scientists participated. The goal of 16th global climate conference was to reach a new agreement to advance collective efforts to limit the global warming to an increase of 2 degrees Celsius above pre-industrial level. The most of the scientists agree that this is the maximum increase, which is necessary to avert the worst consequences of global warming/ climate change, extreme weather pattern, droughts and floods, heat waves and wild fires, melting of glaciers and polar ice caps, sea level rise, oceans acidification, coral bleaching, migration of tropical diseases towards the temperature latitudes and extinction of species. The two weeks Cancun Climate Conference posed number of difficult challenges. The issue of Kyoto Protocol beyond 2012—whether or not extend—has divided the Parties (nations) in the 16th UN global climate conference.

In the last day of 16th UN global Climate Conference at Cancun the participant parties (nations) or the international community adopted the Cancun Agreements. The Cancun Agreements contain series of decisions under the working group on the Kyoto Protocol and the working group on long-term cooperative action [AWGLCA] prepared by the Mexican presidency and based on the two weeks extensive consultations. The negotiators in Cancun (Mexico) climate conference have in the last few hours approved the "Cancun Agreement" in spite of objections from the Bolivian delegation to reach a deal on a wide-ranging package of measures to curb global green house gas emissions, improve forest protection and set up a new global "green climate fund". The Mexican foreign secretary Patricia Espinosa banged down her gavel to formally adopt two draft negotiating texts on the future of the Kyoto Protocol and the parallel long-term cooperative action (LCA) negotiating strand. The texts were adopted despite objections from the Bolivian delegation. Although some groups continued to feel that the process was not transparent and the expression of civil society was constrained. When the final texts were presented to the formal meeting the chair met with rapturous applause. The Bolivia raised objections to the texts, pointed out the inadequacy of its commitments to protect vulnerable peoples from devastating impacts of climate change. It looked like there could be a reopening of the text. Saudi Arabia also saw an opportunity to express its concerns. This was followed by an extensive round of interventions from developed and developing countries alike, including many of the most vulnerable countries: India, China, European Union, US and Japan. All stated that while the text is not perfect but they are satisfied that it represents both progress and broad consensus. They expressed about the delicate nature of the package of compromises and cautioned against reopening of text and emphasized that the next year would see renewed negotiations to build further on the consensus agreed in Cancun Climate Conference. After the adoption of the documents, Norway said that many in the participants of Cancun conference shared

Bolivia's concerns but these concerns could be addressed in years to come and Cancun Agreement was not the final agreement but a major step in the right direction. The Bolivian delegation sharply reacted over the COP 16 president's gaveling of the decisions as being adopted despite the objection, giving rise to argument as to what the meaning of consensus. In the practice of the UNFCCC the decisions have been taken on the basis of consensus. The Bolivia's ambassador Solon said that the consensus required the absence of explicit rejection or objection to a decision proposed for adoption and that consensus did not mean a majority being in favour to adopt a decision. In contrast, the Mexico's foreign minister Espinosa said, 'consensus did not mean unanimity or a right of a party to veto a decision. US climate envoy Todd Stern who supported the adoption of the agreements said that the practice in this body has been general agreement rather than consensus. Many countries both developed and developing expressed support for the two texts and said, they reflect balance, pragmatism and compromises, while many used qualifications such as that they were imperfect or that they felt guarded optimism. The few countries, particularly the Cuba and Ecuador raised their concerns over several issues in the texts and many countries said that following the failure to reach agreement in Copenhagen, the Cancun outcome restored confidence in the multilateral process. The other countries were even more upbeat. Australia called the agreements a game changing moment for the climate change. The feeling among many delegations in the Cancun Climate Conference was that the texts should be adopted in order to avoid the collapse of another climate conference following the failure of the Copenhagen Climate Conference. Several delegates having the opinion that another failure would further dent the image of the UNFCCC and multilateralism and it would be difficult for the talks to regain momentum. In this atmosphere many developing countries decided to go along with the drafts even though they had concerns on various parts. In fact, the Cancun Agreement reached in the tune of Copenhagen Accord. In Cancun the Copenhagen Accord was legitimized and advanced each of the core elements of the Copenhagen Accords. Indeed the Cancun Agreements are not perfect but there are important elements of progress in Cancun Agreement. This includes the recognition within the decisions both under the Kyoto Protocol and the long-term cooperative action track, the need for an increase in current emission reduction pledges to bring them into line with the levels recommended by the UNIPCC to avoid dangerous climate change. For developed countries this will mean the raising of their emission reduction pledges from their existing 17% to between 25 and 40% based on 1990 levels by 2020. But there is also demand for the reductions of more than 40% to provide a reasonable chance to achieving the goal. The important inclusion in the Cancun Agreements was the agreement on the need for periodical review of the adequacy of the agreed goal for the limitation of a further rise in the earth's surface temperature, now agreed as 2 degrees centigrade. This keeps open the possibility of lowering this to the safer limit of 1.5° C as the loud demand of the majority of developing countries. There was also agreement on the establishment of an adaptation framework, including an expert committee to improve global coordination of adaptation efforts. The important achievement of Cancun Climate Conference was the reinvigoration of the multilateral process. The Indian environment minister Mr Jai Ram Ramesh stated that the Cancun Climate Conference had given a signal of hope for multilateralism at a time when this was badly needed.

The perspective of sustainable development, justice and adaptation to climate change demands deep and urgent reductions in GHG emissions from developed and major developing countries. The adaptation to climate change and to pursue the sustainable development, particularly by the poor and developing countries need a fair and legally binding agreement. But the international community is still far off from the real goal. The US envoy in Cancun Climate Conference Mr Stern said that throughout the year (from Copenhagen Climate Conference to Cancun Climate Conference) our strategic vision was to consolidate and elaborate on the progress made in Copenhagen by many of the world leaders led by US president Obama and to such outcome fully endorsed by the conference of the parties, all the nations to the climate treaty, as the Copenhagen Accord was not. The Cancun Climate Conference endorsed the non-binding deals reached at 15th global climate conference at Copenhagen. There was no agreement on extending the Kyoto Protocol beyond 2012 and there was no commitment to continue the Kyoto Protocol after it expires in 2012. The Kyoto Protocol requires 37 signatory countries to reduce emissions an average of 5.2 percent below 1990 levels by the end of 2012. Japan, Russia, Canada and Australia—all bound by the Kyoto Protocol—have indicated that they are not willing to extend the treaty. Many vulnerable developing countries raised their voice that the world must retain the only treaty that puts hard limits on green house gas emissions. Mr Stern said that the hesitance on the part of some countries to want to go into a second period, given that a second Kyoto period would probably only cover 20% of global emissions. This does not have the US in it and you don't have any commitment from the major developing countries. The Kyoto Protocol is not the larger agreement that includes the emission cut commitments from the US, China, India, Brazil, South Africa, etc.

In fact, the Cancun text is a deviation and side track from the decisions taken in Earth Summit, Kyoto Protocol and Bali Action Plan. This will eventually lead to the death of Kyoto Protocol, the only treaty that imposes legally binding green house gas cuts on developed countries. The developing countries will be subjected to the international verification only if their emission curbing mechanism will be funded by the developed countries. The developed rich countries will report on emission of GHG cuts against the international standards but the poor nations must also report on GHG emission cuts against international standards, only after funding has been delivered to build the appropriate mechanism. The financial promises that were made in the Copenhagen have continued in the Cancun agreements. It is the core part of the deal. These include fast-start funding of US $ 30 billion from industrial countries to help the vulnerable countries to cope with the threat of global warming/climate change. The rich developed countries agreed to jointly come up with US $100 billion annually by 2020 to mitigate the impacts of climate change on developing countries that are emitting comparatively very little green house gases. But the Bolivia said that the text states that US $ 100 billion per year would be mobilized by 2020 but does not state clearly whether this will come from developed countries or from the carbon markets. The green climate fund to be established, the Bolivian delegate Solon said, "..well designed fund". But it cannot be effective without funds and the text invites the World Bank to serve as the interim trustee. The Bolivia could not accept this as the World Bank is an institution dominated by the donors.

Mr Stern explained that the Cancun agreement outlines a system of transparency with substantial detail and content, including international consultations and analysis. This will provide confidence that the countries pledges are being carried out and will help the world to keep track in terms of the reduction of the green house gases emission.

The Cancun Climate Conference agreement texts were adopted despite objections from the Bolivian delegation, which expressed its opposition to both the texts. The Bolivian delegation warned that the targets contain in the documents would condemn the world to average temperature increases of over 4 degrees centigrade and stated that, "if the Cancun text was a small step forward, would support it with reservations but in fact, this is a step backwards". Solon elaborated on Bolivia's opposition. He said that parties did not mandate anyone to produce the document and that text was the negotiating text which contained parties positions. Bolivia could not support the 2°C goal as according to various studies, this would mean a 3°C situation for Africa. Referring to the UNIPCC fourth assessment report, a 2°C target would mean a 50% chance of stabilizing the climate. We did not come here and have a blank cheque where the annex 1 parties fill up whatever they want which is not related to the shared vision. When the draft decision on the Kyoto Protocol process under the ad hoc working group on Kyoto Protocol was presented by Espinosa for adoption under the CMP, Bolivia gave its reasons that why it was opposed to the decision. Solon said that this decision did not represent a step forward but was a step backwards as what was being done to postpone indefinitely a decision under the Kyoto Protocol and it opened the doors to a regime which will be flexible and voluntary for a pledge and review approach and not a system where all annex 1 parties will fulfil a set target. Solon by referring paragraph 36 of the text where parties took note of quantified economy wide emission reduction targets to be implemented by annex 1 parties as communicated by them and contained in document, said that this document did not exist and that parties do not know what these commitments will lead to a stabilization of green house gas emissions which will be sustainable for human life and plants. He added that if what was being referred to in the document were the Copenhagen Accord pledges, this would only amount to a 13% to 17% reductions in emissions compared to 1990 levels, which will lead to a temperature rise of 4°C. Solon said that such a temperature level could lead to a situation of genocide and ecocide. Bolivia could not agree to an agreement which will put more human lives in a situation close to death. Bolivia had come to Cancun to ensure that there would be a decision to guarantee a second commitment period under the Kyoto Protocol and this document did not guarantee that. Solon at the plenary said that his government wanted modifications to the text, which was asked to be adopted without any modification or amendment. If the document represented a step forward, we would have supported it. In reality, what is going to emerge is not a stronger regime for reducing emissions in mitigation but a voluntary regime which is less demanding on developed countries that are responsible for global warming/climate change. On technology transfer, Solon said that the new technology executive committee to be established is not even able to address the intellectual property right issues. He said that in most clean technologies 70%–80% of that is in the hands of developed countries and it is they who hold the patents. The Bolivian delegate said that the Cancun documents did not include its proposals. We represent a small country which has principles and will not sell our country and we will speak with the people of the world. There is no consensus for approval of this document. Espinosa in response said that the text was the result of collective work and the concerns of Bolivia would be reflected on the record. She gavelled and announced the adoption of the texts. Solon again took the floor and said that consensus meant that no state was explicitly stating objection or rejection to the decision on. He added, you cannot say there is consensus. You can only take note. This is an attempt to violate rules of the convention and the United Nations. Consensus is not by a majority. There must be an absence of explicit rejection of a decision. Despite

Bolivia's objection, the decision was adopted. Solon declared that we are going to apply to all international bodies to preserve the rules of consensus. We came here to negotiate and not to gavel an injustice. Not even in Copenhagen was this done and there was respect for the rule of consensus. Now there is a violation of rule and this is an unhappy conclusion. He said, I ask you to review your decision and return to the path of law. In response, Espinosa said that on the matter of procedure, consensus did not mean unanimity. At this point, she said that we could not disregard the opinion of 193 parties to adopt the decision, which had been duly adopted.

Under the UN rules the unanimous support is required for agreements to be formally reached. The US negotiator Mr Todd Stern told in the final plenary session that the meeting should adopt procedures that are closer to general agreement the consensus. The Mexican President Felipe Calderon praised negotiators for delivering an historic deal, noting that a good agreement is one where everyone is a little dissatisfied. The delegates and many green groups broadly welcomed the compromise agreement and they expressed that it represents "an important step" towards a formal treaty being conference finalized in South Africa next year. However, all parties in the Cancun UN climate acknowledged that they are still a long way from a final agreement with both texts effectively delaying decisions on the most contentious issues, such as the future of the Kyoto Protocol and the need to close the so-called "gig atone gap" by agreeing more ambitious emissions targets for 2020 that are in line with scientists recommendations. The LCA text, which runs parallel to the Kyoto Protocol negotiations and includes the US, sets a target of limiting temperature rises to 2 degrees centigrade and concession to island countries promises to review whether the goal should be lowered to 1.5 degrees centigrade. It also brings the voluntary emission targets put forward by nations as part of Copenhagen Accord into the UN negotiating process for the first time. It also proposes the formation of New Green Climate Fund, to manage the distribution of funding to help developing countries to reduce the emission of GHG and adapt to climate change. The fund will be managed by the UN, not by the World Bank. The World Bank will only serve as a trustee to the fund. The board of the New Green Climate Fund will have 25 members from developing countries and 15 members from industrialized developed countries.

The Cancun agreement text provides detailed proposals on how nations should monitor the reports and verify their performance against GHG emission cut targets. This was important issue in Cancun talks that has been insisted upon by the US but it was strongly opposed by the China.

In Cancun Climate Conference many of the big questions remained unanswered and pathway forward is still unclear. The Kyoto text effectively delays the final decision on second commitment period for the Kyoto Protocol for next year and left the most contentious issue at the Cancun climate summit unresolved. There is speculation that the European Union could try to negotiate a compromise next year between Japan and Russia, who have said that they will not agree to an extension of Kyoto Protocol. The developing countries who are insisting second commitment period by agreeing two year extension to Kyoto Protocol while a legally-binding replacement is finalised. The Cancun Agreements have kept open the possibility of a second commitment period of the Kyoto Protocol. Similarly, under the decisions of long-term cooperative action kept the option of a legally binding outcome open but did not decide definitely that this is the ultimate objective of continued negotiations.

The deputy chief executive of the climate group think tank Mr Mark Kenber hailed the Cancun agreement as a major shot in the arm to the international climate process. He said the Mexican government has led an impressive balancing act to rebuild trust and revive multilateral negotiations which were in danger of permanent deadlock. The countries have recognised that tackling climate change is good for both the economies and the environment. They are at the forefront of a clean industrial revolution and the Cancun Climate Conference has sent a strong signal to the world that this revolution is underway. The organization named the Friends of The Earth offered the deal a less ringing endorsement, describing it as a weak and ineffective agreement. The US, Russia and Japan backed by powerful vested interests have pursued the agenda which has opened the door to a hazardous system where GHG emissions cut targets would be decided on the whim of politicians, rather than by science.

The US climate negotiator Todd Stern said, "From our point of view what just happened is really very significant. It lays out the structure of an international agreement in all of the crucial areas. This package of agreements obviously is not going to solve climate change by itself but it is a very good step and a step that is very much consistent with US interests and will help to move the world down a path toward a broader global response to stopping climate change. The US is pleased that parties showed the flexibility and pragmatism that was necessary to make progress in each of these areas. Mr Todd Stern said on Kyoto Protocol track that there may be some kind of legal treaty in the future but that is not happening soon because the reason is that the US is not prepared to enter into legally binding commitments to reduce GHG emissions unless China, India and others are also prepared to that. The US position on China is that China needs to make significant reductions in its emissions. But the China and other developing countries at this stage are going for relative reductions. The movement of emissions is very much linked to the movement of economic growth. That's why we are not calling on China, India and other major developing countries to make legally binding commitments right now. The US will do legally binding commitments only if the emerging market countries do that also. The critical direction we need to move on is to separate growth from the path of GHG emission cuts, so that the growth goes up but GHG emissions go down.

Australia said the package of decisions adopted in Cancun Climate Conference was a game changing moment for the climate regime. Pakistan said that the Cancun agreement reflected pragmatism and hope. In fact, it cannot satisfy all parties but it was no doubt a decisive step to the future. It also stressed the need for a more solid outcome by adopting a second commitment period for annex 1 parties under the Kyoto Protocol. Nicaragua said that it was important to make last efforts to hear positions of all Parties with flexibility and mutual understanding. Lesotho for the less developed countries, said that the Cancun package laid a good foundation for further work at the next COP 17 in Durban, South Africa. Cuba, referring to the stand of Bolivia, said that it represents the demands of the indigenous people of the America. It said that the Bolivia speaks on behalf of movements in Latin America and therefore deserves attention. Cuba expressed dissatisfaction with the text of the AWGLCA. It said that the text did not have clear-cut GHG emission reduction goals and it is key to have the second commitment period under the Kyoto Protocol. India's environment minister Mr Jairam Ramesh said that Espinosa had restored confidence in the multilateral process at a time when confidence was at an historical low. He said that parties could confidently look ahead and approach the challenge of climate change in a spirit of constructive compromise. Singapore said the text was not perfect but in any negotiations, not everyone will get what they want. It said that there were some missing elements that would need to be clarified. It said the document was not but a step forward for a legally binding agreement in next COP 17 in Durban. The 17th Global Conference on the challenge of Global Climate change held at Durbin, South Africa December 2011 under UNIPCC. Thousands of delegates all over the world representing the national Governments, Environmentalist, Social activists and NGOs participated in the Durbin Conference. In spite of some agreements the Durbin Global Climate Conference like previous conferences could not achieve the main goal of concluding the legally binding Global Treaty. The European Union climate commissioner Connie Hedegaard said, "that multilateralism had shown results". The EU came to Cancun hoping for a balanced package that has been achieved." China represented by Minister XieZhenhua said that parties had demonstrated good political will for a balanced result. He said that the text provided a sound basis for future negotiations. There were shortcomings but it was satisfied that the negotiations had adhered to the Bali Roadmap. Ecuador said that it did not like all the results and said that parties must redouble their efforts to ensure the health of the Planet Earth. It stressed the need for the second commitment period of the Kyoto Protocol to be confirmed and to come into effect after 2012. It supported the observations of Bolivia on the various deficiencies in the text and said that it must be improved. Brazil said that the document was balanced in essence although not perfect and there was a sense of realism and pragmatism.

In Cancun parallel to official UN global climate conference many civil society groups and social movements have gathered and organized debates about the state of the negotiations and alternative policy approaches. They criticized the inadequacy of Cancun Outcome and Highlighted the need for a fundamental shift in production and consumption models.

WHO IS RESPONSIBLE

The corporate capitalist institutions in the world: the UN, the World Bank, IMF, WTO, nation-states and the prominent scientists, thinkers and planners have common view that the fundamental crisis/challenge of global warming/climate change is the outcome of human activity. But nobody ever defined the concept of human activity. The global summits on global warming and sustainable development also have not touched this basic question. While adopting the agenda of sustainable development they have made only a passing reference to the unsustainable human ways of life or lifestyle, but shirked to pinpoint the specific structural form (social system) in which mankind, in given conditions, operates in various modes. The global corporate capitalist forces have left this basic and essential point quite vague by putting an equal blame on every human being. According to this standpoint that every human being is a polluter is not correct but in reality the responsibility of producing pollution or green house gases rests with those who own pollutant producing factories, farms and different development sectors (the corporate capitalists and big landlords) not the workers or common people of the world. The historical experience shows that the way of life or life style or human activity has always existed in an organized social system. Today mankind is living and doing activities within the global corporate capitalist system which is functioning through multiple nation-states. The historical facts prove that the existing global corporate capitalist system and complex multi-natural factors are responsible for the fundamental challenge/crisis of global warming/climate change.

REPAIRING OF ENVIRONMENTAL DEGRADATION

The damage of environment and climate system of Planet Earth will take more than one century to repair. The repairing project of environment and climate system needs trillions of money. The question arises from where this huge amount of money will come. The law of rationality demands that, the most industrialized developed countries who damaged the environment more, are liable or morally bound to take the more share of responsibility of damage repair. It will be more appropriate that the task of Planet Earth's environment repairing and generation of alternative clean energy be handed over to the UN general assembly or nature-mankind friendly new global democratic confederal system democratically elected by the nations of the world or world people, instead of entrusting it over to the UN security council or G 8 group or big global corporate capitalist powers. All are dominated by the global corporate capitalist vested interests (the real creator of global warming/climate change and global dehumanization/human-inequality). In the UN general assembly where all the decisions are taken on the majority bases, there is little likelihood of the domination of super powers or rich nations.

SCEPTIC VIEWPOINT ABOUT GLOBAL CLIMATE CHANGE

The UN IPCC should adopt democratic style (a style to listen dissent or minority viewpoints also). The minority viewpoint or sceptic viewpoint about global warming/climate change should also be provided proper space for expression and deliberation. The democratic existence of the majority and minority viewpoints boosts the comparative scientific investigation, study and research and provides comparative knowledge to the mankind or human society. The minority of professional associations and individual scientists or group of some scientists have issued non-committal statements. When the UNIPCC report came out, physicists Mr Seitz, Mr Nierenberg and Mr Fred Singer criticized the scientist Mr Santer. In an open letter to the UNIPCC, Mr Seitz, Mr Nierenberg and Mr Fred Singer accused Mr Santer of making unauthorized changes to the UNIPCC report.

The scientists opposing the main stream consensus scientific assessment of global warming/climate change express varied opinions concerning the cause of global warming. The scientists Mr Balling, Mr Lindzen and Mr Spencer say that it has not yet been ascertained whether humans are the primary cause of the global warming/climate change. The scientists Mr Soon, Mr Balimuos, Mr Grey,

Mr Viezer and Mr Leroux holds that the global warming is the result of natural variations, oceans currents, increased solar activity, cosmic rays or unknown natural causes. A few studies claimed that the present level of solar activity is historically high as determined by the sun spot activity and other factors. The solar activity effect on climate either by variations in the sun's output or more speculatively by indirect effect on the amount of cloud formation.

The Non-Governmental International Panel on Climate Change (NIPCC) was setup to examine the climate data used by the UN sponsored International Panel on Climate Change (UNIPCC). The Non-Governmental International Panel on Climate Change (NIPCC) issued a report under title "Climate Change Reconsidered 2009". In this report, it is mentioned that the UNIPCC released its fourth assessment report in 2007 titled "Climate Change 2007". The UNIPCC claimed that it is most comprehensive and up to date report available on the subject and constitute the standard reference for all concerned with climate change in academia, government and industries worldwide. The most important issue the UNIPCC claimed that most of the observed increase in global average temperatures since the mid-century is very likely due to the observed increase in anthropogenic green house gas concentrations. But NIPCC reaches the opposite conclusion that the natural causes are very likely to be the dominant cause of global warming/climate change. It doesn't say that the anthropogenic green house gases cannot produce some warming or has not in the past. Its conclusion is that the evidence shows they are not playing substantial role.

The UNIPCC says that the global climate change will increase the people suffering from death, diseases and injury from heat waves, floods, storms, fires and droughts. The NIPCC again reaches the opposite conclusion: A warmer world will be safer and healthier for mankind and wildlife. Once again, we do not say that global warming/climate change would not occur or have any affects (positive or negative) on human health and wildlife. Rather our conclusion is that evidence shows that net effect of continued warming and rising carbon dioxide concentration in the atmosphere will be beneficial to humans, plants and wildlife.

The NIPCC reviewed the material presented in the UNIPCC fourth assessment report: the physical science basis and impacts, adaptation and vulnerability. It finds them to be highly selective and controversial with regard to making future projections of climate change and discerning a significant human-induced influence on current and past climatic trends. Although the UNIPCC claims to be unbiased and to have based UNIPCC fourth assessment report on the best available science. But in many instances the conclusions have been seriously exaggerated, relevant facts have been distorted and key scientific studies have been omitted or ignored. The NIPCC contradicted the UNIPCC's central claim that global warming/climate change is man-made and its effects will be catastrophic. The rise in environmental consciousness since 1970 has focused on a succession of calamities, cancer epidemics from chemicals, extinction of birds and other species by pesticides, the depletion of ozone layer by supersonic transports and later by Freon, the death of forests because of acid rain and finally global warming/climate change.

The UNIPCC can trace its roots to World Earth Day in 1970, the Stockholm conference in 1971–72 and the Villach conference in 1980 and 1985. The UN environmental progress in 1986 and the World Meteorological Organization established the Intergovernmental Panel on climate change as an organ on climate change. The UNIPCC key personnel and lead authors were appointed by the governments and its summaries for policy maker have been subjected to approval by the member government of the UN. The scientists involved with the UNIPCC are almost supported by the government contracts which pay not only for their researches but for their IPCC activities.

The UNIPCC agenda is to justify the control of emission of green house gases (GHG), especially carbon dioxide. Consequently, its scientific reports have focused solely on evidence that might point towards human-induces climate change. The role of UNIPCC is to assess on a comprehensive, objective, open and transparent basis the latest scientific, technical and socio-economic knowledge produced world-wide, relevant to the understandings of the risk of the human-induced climate change, it's observed and projected impacts and options for adaptation and migration.

The UNIPCC's three chief ideologists have been Prof. Bert Bolin, a meteorologist at Stockholm University, Dr Robert Watson, an atmospheric chemist at NASA, later at the World Bank and new chief scientist at the UK department of environment, food and rural affairs and Dr John Houghton, an atmospheric radiation physicist at Oxford university, later the head of the UK Met office as Sir John Houghton.

The UNIPCC from the very beginning a body of leading scientists reflecting the position of their governments or seeking to induced their governments to adopt the UNIPCC position. A small group of activists wrote the all important summary for the policy makers for each of the four IPCC reports. It was told that the thousand of scientists on whose work the assessment reports are based but the vast majority of the scientist has no direct influence on the conclusion expressed by the IPCC. The policy summaries were produced by the inner core of scientist and were reviewed and agreed to line by line by the representatives of the member governments.

The UNIPCC first assessment report concluded that the observed temperature changes were broadly consistent with green house gas model. Without much analysis, it gave the climate sensitivity of a 1.5 to 4.5 degrees Celsius rise for a doubling of green house gases. The UNIPCC second assessment report 2006 contained the memorable conclusion, "the balance of evidence suggests a discernible human influence on global climate". The second assessment report was again heavily criticized. This time for undergone significant changes in the body of the report to make it conforms to the summary for policy makers after it was finally approved by the scientist involved in writing the report. The report was not only altered but key graph was also doctored to suggest a human influence. The evidence presented to support the SPM conclusion turned out to be completely spurious. This led to heaved between the supporters of the UNIPCC and those who were aware of the altered text and the graph. The UNIPCC second assessment report also provoked the 1996 publications of the Leipzig declaration by Science and Environment Policy Project (SEPP) which was signed by 100 climate scientists in the form of booklet titled "The Scientist Case Against the Global Climate Treaty". The IPCC report, in spite of obvious shortcomings, provided the under pinning of the Kyoto protocol. The background is described in the detail in the booklet "Climate policy from Rio to Kyoto" published by Hoover institution 2000.

The UNIPCC third assessment report was noteworthy for its use of spurious scientific papers to backup its summary for its policy makers claim of new and strong evidence of anthropogenic global warming. One of these was so-called hockey stick paper, an analysis for proxy data, which claimed the twentieth century was the warmest in the past 1000 years. The paper was later found to contain errors in the statistical analysis. The IPCC also supported a paper that claimed pre-1940 warming was of human origin and caused by green house gases. This work too contained fundamental errors in its statistical analysis. The Science and Environmental Policy Project has given its reactions about the third assessment report in the booklet titled, The Kyoto Protocol is not Backed by Science 2002.

The UNIPCC fourth assessment report was published in 2007; the summary of working group 1 was released in February 2007 and the full report in May 2007, after it had been changed once again to conform to the summary. It is significant that the IPCC fourth assessment report; no longer make use of the hockey stick paper or the paper claiming the pre-1940 human caused global warming. Once again controversies ensued, when the IPCC refused to publicly share the comments submitted by the peer-reviewers but under pressure posted them online. The inspection of those comments revealed that the authors had rejected more than half of all the reviewers' comments in the crucial chapter attributing the recent global warming/climate change to human activities. The fourth assessment report concluded that the most of the observed increase in global average temperature since the mid-20th century is very likely due to the observed increase in anthropogenic green house gases concentration. However, it ignored the available evidence against the human contribution to current global warming/climate change and the substantial research of the past few years on the effects of the solar activity on climate change.

Keeping in view the errors and falsehood observed in the assessment reports of the UNIPCC regarding global warming/

climate change the Science and Environment Policy Project set up a team B in 2003 to produce an independent evaluation of the available scientific evidences. The name of team B was changed into Non-Governmental International Panel on Climate Change (NIPCC). The NIPCC is an international panel of non-governmental scientists and scholars who have come together to understand the causes and consequences of climate change. NIPCC organized an international climate workshop in Vienna in April 2007. The Climate Change Reconsidered, a report of NIPCC was the result of Vienna workshop and subsequent research and contributions by a larger group of international scholars.

NIPCC Report—a Brief

The forecasts in the fourth assessment report were not the outcome of validated scientific procedures. The UNIPCCs claim that it is making projections rather than forecasts is not a plausible defence. Today's climate models fail to accurately simulate the physics of earth's radiative energy balance resulting in uncertainties, as large as the doubled carbon dioxide forcing.

The scientific research suggests the model derived temperature sensitivity of the earth accepted by the UNIPCC is too large. The corrected feedbacks in the climate system could reduce climate sensitivity. The scientists may have discovered a connection between the cloud creation and the sea surface temperature in the tropics that creates a "thermostat like control" that automatically vents excess heat into space. It conformed this could totally compensate for the warming influence of all the anthropogenic carbon dioxide emissions. The UNIPCC under estimated the cooling effect of aerosols. The radiative effect of aerosols is comparable larger than the temperature forcing, caused by all the increase in the green house gases concentrations. The high temperature is known to increase emission of dimethyl sulphide from the ocean which increases the albedo of marine stratus clouds, which has cooling effect. Iodo compounds created by marine algae functions as cloud condensation nuclei which help to create new clouds that reflect more incoming solar radiations back to the space and thereby cool the Planet Earth. As the air 0.5 carbon dioxide content and its temperature continue to rise. The plants emit greater amount of carbonyl sulphide gas which eventually make its way into the stratosphere where it is transformed into solar radiation reflecting sulphate aerosol particles which have a cooling effect.

The carbon dioxide enrichment enhances the biological growth and atmospheric levels of bio-sols rise, many of which functions as a cloud condensation nuclei. The increase cloudiness diffuses light, which stimulate plant growth and transform more fixed carbon into plant and soil storage reservoirs. The agriculture accounts for the almost half of the nitrous oxide in some countries. There is concern that enhanced plant growth due to carbon dioxide enrichment might increase the amount and warming effect of the green house gases. But the field research shows that nitrous oxide emission fall as carbon dioxide concentration and temperature rises.

The UNIPCC claims to find evidence in temperature records that the warming of the 20th century was unprecedented and more rapid than during any previous period in past 1300 years. But the evidence it cites including the "hockey-stick" representation of earth's temperature recorded by Mann et al. has been discredited and contradicted by many independent scholars. A corrected temperature record shows temperature around the world were warmer during the Medieval Warm Period of approximately 1000 years ago than they are today and have average 2–3 degrees Fahrenheit warmer than today's temperature over the past 10,000 years. The evidences of the Medieval Warm Period are extensive and irrefutable. The highly accurate satellite data adjusted from orbit drift and other factors show much more modest warming trend in the last two decades of the twentieth century and a dramatic decline in the warming trends in the first decade of the 21st century. The fingerprint or patterns of warming observed in the 20th century differ from the pattern predicted by the global climate model designed to simulate carbon dioxide global warming. All green house models show an increasing warming trend with altitude in the tropics however the temperature data from balloon gives the opposite result – no increase in warming but rather a slight cooling with altitude. The temperature records in Greenland and other Arctic areas reveal that temperature reached a maximum around 1930 and have decreased in recent decades. The long-term studies depict oscillatory cooling since the climatic optimum of mid-Holocene when it was perhaps 2.5 degrees Celsius warmer than it is now. The average temperature history of Antarctica provides no evidence of 20th century warming. While the Antarctica Peninsula shows recent warming; while several research teams have observed cooling trends in the inferior of the Antarctica continent since 1970.

The glaciers around the world are continuously advancing and retreating with a general pattern of retreat since the end of the Little Ice Age. There is no evidence of an increase rate of melting overall since carbon dioxide level arises above their pre-industrial level. Carbon dioxide is not responsible for the melting of glaciers. The ice area of ocean has continued to increase around Antarctica over the past few decades. The evidence shows that much of the reported thinning of Arctic sea ice that occurred in the 1990 was a natural consequence of changes in ice dynamics caused by an atmospheric regime shift. The global studies of precipitations trends show no net increase and no consequent trend with carbon dioxide contradicting climate model predictions that global warming should cause increased precipitation. The research on Africa, Arctic, Asia, Europe, North and South America finds no evidence of a significant impact on precipitation that could be attributed to anthropogenic global warming.

The cumulative discharge of the world's rivers remained statistically unchanged between 1951 and 2000; a finding that contradicts computers forecast that a warmer world would cause large change in global stream flow characteristics. The results of the science research studies are strongly against the claims that carbon dioxide induced global warming would cause catastrophic disintegration of the Greenland and Antarctic ice sheet. The mean rate of global sea level rise has not accelerated over the recent past. The determinants of the sea level are poorly understood due to considerable uncertainty associated with a number of basic parameters that are related to the water balance of the world's oceans and the melt water contribution of Greenland and Antarctica. Until these uncertainties are satisfactorily resolved, we cannot be confident that short-lived changes in the global temperature produce corresponding changes in sea level.

The UNIPCC claims the radiative forcing due to changes in the solar output since 1750 is + 0.12 wm^2 an order of magnitude smaller than its estimated net anthropogenic forcing of + 1.66 wm^2. The research suggest that the IPCC has get it backwards, that it is the sun's influence that is responsible for the lion's shares of the climate change during the past century and beyond. The total energy output of the sun changes by only 0.1% during the course of the solar cycle. Although the larger may be possible over periods of centuries. But the ultraviolet radiations from the sun can change by several percent over the solar cycle—as observing changes in stratospheric ozone. The largest changes occur in the intensity of the solar wind and inter planetary magnetic field. The ancient climates reveal a close correlation between solar magnetic activities and solar irradiance (brightness) on the one hand and temperature on earth on the other hand. The correlation is much closer than the relationship between carbon dioxide and temperature. The cosmic rays could provide the mechanism by which changes in solar activity affect climate. During the periods of greater solar magnetic activity, the greater shielding if the earth occurs resulting in the less cosmic rays penetrating to the lower atmosphere and produces fewer cloud condensation nuclei. This creates fewer and less reflective low-level clouds which lead to more solar radiations observed by the surface of the earth, resulting in increasing near surface temperature and global warming. There is strong correlation between solar variability and precipitation, droughts, floods and monsoons. The correlations are much stronger than any relationship between weather phenomenon and carbon dioxide. But in fact the role of solar activity in causing climate change is so complex that most theories of the solar forcing must be considered to be as yet unproven. But it would also be appropriate for climate scientist to admit the some about the role of the rising atmospheric carbon dioxide concentration in driving recent global warming/climate change.

The UNIPCC predicts that a warmer Planet Earth leads to more extreme weather, characterized by the more frequent and serve episodes of droughts, flooding, cyclones and precipitation variability, storms, snow, storm surges, temperature variability and wild fires. But has the last century during which the IPCC claims the world experience the more rapid warming than any time in the past two millennia experienced significant trends in any of these extreme weather events. The droughts have not become more extreme or erratic in response to global warming. The floods were more frequent and more severe during the Little Ice Age than they have been during the current warm period. The flooding in Asia, Europe and North America has tended to be less frequent and less severe during the 20th century. The UNIPCC says that, it is likely that the future tropical cyclones (typhoons and hurricanes) will become more intense, with larger peak wind speeds and heavier precipitation associated with ongoing increase of tropical sea surface temperature. Despite the supposedly unprecedented warming of the 20th century, there has been no increase in the intensity or frequency of the tropical cyclones globally or in any of the specific ocean.

The Planet Earth has warmed over the past 150 years during its recovery from the global cool of the Little Ice Age. There has been no significant increase in either the frequency or the intensity of the stormy weather. The storm surges have not increased in either frequency or magnitude as carbon dioxide concentration in the atmosphere has risen. The 300 ppm increase of carbon dioxide in the air content typically raises the productivity of most herbaceous plants by about one-third and this positive response occurs in plants that utilizes all three of the major bio-chemical pathways of the photosynthesis. The productivity benefits of carbon dioxide enrichment are also experienced by the aquatic plants including fresh water algae and macrophytes and marine micro-algae and macro-algae. The atmospheric carbon dioxide enrichment helps ameliorate the detrimental effect of several environmental stresses on plant growth and development including high soil salinity, high air temperature, low light intensity, and low levels of soil fertility.

As the carbon dioxide content continue to rise, the plants will likely exhibit enhance rate of photosynthesis and bio-mass production that will not be diminished by any global warming that might occur concurrently. The other biological effects of carbon dioxide enhancement include enhanced plant nitrogen use efficiency, longer residence time of carbon in the soil and increased population of the earthworms and soil nematodes. The aerial fertilization affect of the ongoing rise in the air's carbon dioxide concentration, which greatly enhances the vegetative productivity and its anti-transpiration effect—which enhance plant's water use efficiency and enable plant to grow in that areas that were once too dry for them—are stimulating plant growth across the globe in places that previously were too dry or unfavourable to plant growth, leading to a significant greening of the Planet Earth. The elevated carbon dioxide also reduces and overrides the negative effects of the ozone pollution on plant photosynthesis, growth and yield.

The UNIPCC claims that, new evidences suggest that climate driven extinction is widespread and projected impacts on the bio-diversity are significant. The global losses on bio-diversity is irreversible. These claims are not supported by scientific research. The world's species have proven to be remarkable resilient to climate change. The most of wild species are at least 1 million years old which means they have all been through 100s of climate cycle including temperature change on par with or greater than those experienced in the 20th century. The four known causes of extinctions of species are huge asteroid striking the Planet Earth, human hunting, human agriculture and introduction of alien species. None of these causes are connected with global temperatures or atmospheric carbon dioxide concentrations. The UN environment programme shows that the role of species extinctions at the end of the 20th century was the lowest since the sixteenth century despite 150 years of rising temperature, growing population and industrialization. The land animals also tend to migrate pole ward and upward to those areas where cold temperature prevented them from going in the past.

The persistence of coral reef through geologic time-when temperature were as mid 10–15 degrees Celsius warmer than at present and atmospheric carbon dioxide concentration two to seven time higher than they are currently-provides substantive evidence that these marine entities can successfully adopt the dramatically changing environment. The rising carbon dioxide content of the atmosphere may induce very small changes in the well-buffered ocean chemistry, that could slightly reduce the coral calcification rates but potential positive effects of hydrospheric carbon dioxide enrichment may more than compensate for this negative phenomenon. The world observation indicated that the elevated carbon dioxide and the elevated temperature are having a positive effect on most corals.

The UNIPCC argues that the global warming/climate change currently contributed to the global burden of diseases and pre-mature death and increase malnutrition and coregent disorders. In fact the overwhelming evidence shows that the higher temperature and rising carbon dioxide levels have played an indispensable role in making it possible to feel growing global population without encroaching on natural ecosystem. The global warming reduces the incidence of cardio-vascular diseases related to low temperatures and wintry weather than cardiovascular diseases associated with high temperature and summer heat waves. The historical increase of the air's carbon dioxide content has probably helped to lengthen human life spans since the advent of the industrial revolution. The bio-fuels for the transportation (chiefly ethanol, bio diesel and methanol) are being used in growing quantities in the belief that provide environmental benefits. In fact these benefits are very dubious.

There is a majority scientific consensus on global warming/climate change that it is the result of human activities. This has been endorsed by every national science academy, which has issued statement on global warming/climate change, including the science academies of all major industrialized countries. The national and international scientific bodies have not issued any dissenting statements. The UN IPCC reports, governmental reports, environmental groups, etc. virtually having unanimous agreement on scientific community consensus in support of human caused global warming/climate change. In 1997 the world scientists' call for action petition was presented to the world leaders meeting to negotiate the Kyoto Protocol. The content of the petition was "a broad consensus among the world's climatologists that there is now a human influence on global warming/climate change. It urged the governments to make legally binding commitments to reduce the industrial nation's emissions of heat trapping gases (GHGs) and called global warming/climate change, one of the most serious threats to the Planet Earth and to the future generations. The petition was conceived by union of concerned scientists as a follow up to their 1992 world scientists warning to mankind of the world. It was signed by more than 1500 world's most distinguished senior scientists, including the majority of Nobel laureates in science.

In the year 2001 sixteen of the world's national science academies made a joint statement on global warming/climate change. These science academies gave their support for the UNIPCC process as under, "the work of Intergovernmental Panel for Climate Change (IPCC) represents the consensus of the international scientific community on climate change science. We recognized the UNIPCC as the world's most reliable source of information on global warming/climate change and its causes. We endorse its method of achieving consensus. Despite increasing consensus on the science under pinning predictions of global warming/climate change doubts have been expressed recently about the need to mitigate the risks posed by the global warming/climate change. We don't consider such doubts justified". Many other science academies and scientific organizations supported the conclusion of the UNIPCC. In the year 1995 the UNIPCC concluded that the human effect on global warming/climate is now discernible. The lead author of the key chapter on detection and attribution was a scientist of the Lawrence Livermore National Laboratory named Benjamin J Santer.

The Indian government has published discussion paper which says there is no conclusive scientific evidence to prove that the Himalayan glaciers are melting due to the global warming/climate change or to link the black carbon in the atmosphere with glaciers. We also cannot link retreating glaciers in the Arctic because of the global warming/climate change to those in the Himalayas. The discussion paper titled "Himalayan glaciers: a state of art review of global studies, glacial retreat and climate change" was released by the union environment minister Mr Jai Ram Ramesh. The paper challenged the UNIPCC assessment report that the Himalayan glaciers are rapidly shrinking due to global warming. The discussion papers was prepared by V K Raina, retired official of the Geological Survey of India with the help of different Himalayan study groups (which are covering only 20–30 Himalayan glaciers) says that, although the Himalayan glaciers are shrinking in volume and showing a retreat, have not been exhibited, especially in recent years, an abnormal annual retreat, of the order that some glaciers of Alaska and Greenland are reported. The discussion paper expresses that the glacier was affected by a range of physical features and complex interplay of climate factors and it would be premature to state that they were retreating as a result of periodic climate variations until many centuries of observation were available. If we see the cumulative average of the past 100 years, no glaciers has deviated from that. There is no abnormal retreat. Mr V K Raina said Gangotri glacier is 30 km long and even if we assume it retreats at a rate of 30 meter a year (as the assessment of the UN IPCC) it will still take 1000 years to disappear. Mr V K Raina said that the views expressed in the discussion paper were not those of the Indian government but its publication was intended to stimulate the discussion. Mr Jai Ram Ramesh said that while most Himalayan glaciers were retreating, some glaciers, like Siachen glacier were actually advancing and Gangotri glacier were retreating at a rate lower than before. The Himalayan glaciers feed major rivers flowing through India, China, Pakistan, Bangladesh, Nepal and Myanmar. The overall health of the Himalayan glaciers is poor as the debris cover had reached the alarming proportions. The discussion paper added that there was no conclusive scientific evidence to link this to global warming/climate change. The Indian union environment minister Mr Jai Ram Ramesh said that he wanted the scientists to discuss the report and would bring it to the notice of Dr. R K Pachauri, chief of UN IPCC. The UNIPCC in its 2007 report had warned that the Himalayan glaciers had fast receding and that at the current rate of depletion, there was a high likelihood that Himalayan glaciers may disappear by 2035 or earlier.

Dr R K Pachauri reacted on this discussion paper that I don't understand the logic of this. I am puzzled where this magical science has come from. This is something indefensible. But later on Dr R K Pachauri realised the mistake of UNIPCC report and inadequacy of proper data and study of Himalayan glaciers.

Mr V K Raina admits in his discussion paper that there is lack of available data and the long-term data exists only 20 to 30 Himalayan glaciers and there was only one automated weather station recording climate data in the Himalayas. Mr Vinuta Gopal, campaign manager of Green Peace India said that the time now is not about trying to find conclusive evidence, the time now is for action. Some scientists say that the research and field data is too limited to conclude a direct link. There is no field data to corroborate that the glaciers will disappear in next 20–30 years. The Himalayan range has about 9575 glaciers and we studied about 30 glaciers and whichever we studied we need more detailed data. If we want to study the glacier behaviour, we need to monitor for 8–10 years but we only manage 2 years at the most said Prof. R K Ganjoo, Director, Institute of Himalayan Glaciology, University of Jammu. Dr. Shakeel Ahmed Romshoo, Associate Professor, Department of Geology and Geo-physics, University of Kashmir said, "that although very few glaciers have been studied and data is inadequate but it is evident that the global warming affects glaciers". The environment ministry of India will collaborate with ISRO to undertake a three-year study to map Himalayan glaciers through satellites. The effect of black carbon on Himalayan glaciers will also be studied. The ISRO, MOEF and Snow and Avalanche Study Establishment (SASE) are shortly going to install a few hundred automatic weather stations (AWS) throughout the Himalaya from northwest to northeast to generate real-time data on the mountain weather—an essential parameter to understand the effect of climate on Himalayan glaciers.

2

Challenge of Global Dehumanization/Human cum Social Inequality

The challenge of global dehumanization/human cum social inequality is other important integrated dimension of the fundamental challenge of mankind or nature-mankind related global crisis. The challenge of global dehumanization (means deprivation and reversal of human values and qualities) and human cum social inequality (social, political, economic, cultural and diplomatic inequality) is also a very grave social imbalance and challenge before mankind in the present global era.

The process of historical development of human society/ human civilization tells us that the phenomenon of global dehumanization/human cum social inequality is rooted with its different historical stages and social systems. The global dehumanization/human cum social inequality remains prevalent in the following historical stages and systems of mankind – clan stage, tribal stage, feudal monarchical stage, national industrial capitalist stage and now the global corporate capitalist stage (except the first stage of food gathering). But now in the present global corporate capitalist system, the gap and trend of global dehumanization/human cum social inequality widened manifold.

The western democratic and market regulated corporate capitalist model claims that there is equality, freedom of expression and open democracy. But in fact, this is the most unequal, unjust and unfair social system/social order, where the powerful dominates the weak and the haves ride over the have not with money, might and the privileges. The trend is prevailing in both national and global level. In every country few are more privileged than the overwhelming majority. At the global level five veto power and nuclear superpowers— US, Russia, China, UK, and France–with US at top- along with newly developed nuclear powers India, Pakistan, North Korea, Israel, etc. possess huge stockpile of nuclear weapons, which can smash our Planet Earth into pieces. Why a free and open democratic system/democracy which is meant for people welfare needs the manufacturing of such a huge stock of nuclear weapons? In real sense, this is the inhuman feature of champions of the democracy. The extraordinary expenditure on mass destruction weapons mainly by the nations is estimated by the UN to be three times more than what is required to meet the costs of solving all the world's problems of hunger, malnutrition, basic health, drinking water, sewerage, housing, and so on. They are investing billions of money in the manufacturing and maintaining of nuclear and mass destruction weapons and on big military establishments, comparative to very less expenditure in human welfare, sustainable development, conservation and promotion of environment/nature. The rivalries and friction between existing nation-states that led to armed conflicts in the past are less likely to do so. There are comparatively few burning disputes between nations over borders. But the internal conflicts within the nations in the world are numerous. Today the main danger of war lies in the involvement of outside nation-states or military actors. The nation states with stable economies and relatively equitable distribution of goods and incomes among their people, are likely to be less shaky socially and ones. The dramatic increase in economic and social inequality within as well as between the countries may reduce the chances of peace. The avoidance and control of internal armed violence depends mainly on the effective performance of national governments and their legitimacy in the eyes of the majority of their people. Today no government can take for granted the existence of an unarmed civilian people and not in position to overlook or eliminate internal armed minorities easily. The world is increasingly divided into nation states capable of administrating their territories and citizens effectively and the growing number of nations with national governments ranging from the weak and corrupts to the nonexistent. These nations produce bloody internal struggles and international conflicts, as we have witnessed in central Africa. There is no immediate prospect for lasting improvement in such conflicting or civil war regions and a further weakening of central government in unstable countries or further balkanisation of the world would increase the danger of armed conflicts. The war or internal conflicts in the 21st century are not likely to be as murderous as it was in the 20th century. But the violence, human suffering and loss will remain omnipresent and endemic in a large part of the world. In fact, the balance of war and peace in the 21st century depends not only on devising more effective mechanisms for negotiation and settlement but depends on the true realization of unitedly counter and solve mankind's common and existential challenge of global climate change and global dehumanization and ending of all types of discriminations, injustice, avoidance of military conflicts and peaceful settlement of all the disputes of the world and the establishment of nature-mankind friendly global confederal system by replacing the anti nature-anti mankind global corporate capitalist system.

Table 2.1: Detail of military budgets, military strength and nuclear status of major countries of the world (2008 estimate)

S. no.	Countries	Number of troops	Military budget (bn. $)	World share (%)	No. of nuclear weapons
1.	United States	3,315,400	700	41.5	9,400
2.	China	7,024,000	84.9	5.8	240
3.	France	779,450	65.7	4.5	300
4.	United Kingdom	417,260	65.3	4.5	185
5.	Russia	3,796,100	58.6	4.0	12,000
6.	Germany	683,150	46.8	3.2	–
7.	Japan	296,550	50	3.2	–
8.	Italy	534,350	40.6	2.8	–
9.	Saudi Arabia	234,500	38.2	2.6	–
10.	India	3,862,300	30.0	2.1	60-80
11.	North Korea	5,995,000	–	–	<10
12.	South Korea	4,210,000	28.5	1.9	–
13.	Pakistan	1,464,000	7.8	0.5	70–90
14.	Iran	2,005,000	7.31	0.49	–
15.	Israel	629,150	13.3	0.9	80

Total global military budget is 1464 billion dollars (1.464 trillion dollars)

The UNO which is regulated on the basis of 1945 charter, which gives veto-power and permanent membership of security council to the militarily most powerful five nuclear powers/ countries of the world—US, Russia, Britain, China, and France. The privilege-based rule is totally discriminatory and unjust. It negates the worldwide accepted principle of political equality—the one entity one vote—norm of the universal suffrage and violates the human rights of the nations. The present global reality and nature-mankind friendly global model demand that the UN should be restructured and remodelled on the basis of democratic and rational principles. The UN should be made relevant with the tune and spirit of nature-mankind friendly new vision of present global era. Similarly the same undemocratic and irrational principle and practice prevails in international financial institutions—international monetary fund and world bank—in which the economic weightage determine the voting strength of each member. Thus these institutions are definitely governed by the financially rich and developed countries especially the US having the 26% of the total voting strength. There is also domination of rich nations that prevails in world trade organization. The global controversy between rich and developing countries over agriculture subsidies and demand of level playing field in trade is persisting. The sustainable future of mankind demand that the above international institutions [World Bank, IMF, and WTO] should be made pro nature-pro mankind.

Today the few developed rich nations (having 15% of the world population) control 80% of the world's material and financial resources and leaving 20% resources for 85% people of the world in 130 poor and developing countries throughout the world. Today still 63% global population are relatively poor as per the report of world bank. The inequality and dehumanization exists in the form of rich nations and poor nations, developed and underdeveloped/developing nation, haves and have not (rich and poor), global gender-inequality (between male and female), urban and rural inequality, inequality within the countries throughout the world, racial and minority discrimination, civilization clash , tribal and caste discrimination/clash. In principle and practice the situation in every country is not different in the prevailing corporate capitalist global order, in spite of variations of the 194 countries in the world due to their different historical backgrounds, physical structures and socio-economic conditions/levels. There is great similarity and commonness in the quality of social life, especially among the developing countries. The political process of almost every country (weather developed or developing) remains dominated by money power, muscle power, unhealthy political competitions and primordial thinking, practice and norms used by most of the political parties and election candidates. The economic process in general operates through cut throat competition through market mechanism and with the use of black money also. The moral values/universe and principles degenerated everywhere in the world. The immoral and unprincipled way of life and practice prevail and dominate everywhere.

The gap of global dehumanization/human cum social inequality is increasing at an unprecedented scale. The large scale unemployment and poverty is the rising phenomenon of the present world. The recent global economic recession (2007 to onwards) also increased poverty and unemployment. The military budgets are rising throughout the world—a biggest unproductive expenditure—instead of more spending on social-human development and conservation cum promotion of nature/environment. The lopsided development reflects the wrong priority and disharmony between nature and mankind and within mankind. These imbalances and disharmonies created the dehumanized state of mankind and degradation or serious damage of environment of Planet Earth and developed anti nature-anti mankind character of the present global corporate capitalist system. Today the mankind is living within a global corporate capitalist system, which is operating through multiple nation-states. The inequality/dehumanized state of human civilization is existing in all the countries of the world and they provided full authority and opportunity to corporate capitalist players and their pro-political forces to pursue the agenda of maximum growth (without caring to maintain the harmony between nature and mankind and within mankind) and money maximization (through profits) and power monopolization. The global corporate capitalist system created crude and unprecedented global human-inequality between rich and poor in the history of mankind. The different UN human development reports proved the growing global human inequality, dehumanization, mass poverty, mass hunger and human deprivation. The United Nations Development Programme [UNDP] and Oxford University recently [July 2010] launched a new index to measure poverty levels, which gives a multi-dimensional picture of people living in hardships. The multi-dimensional poverty index captures distinct and broader aspects of poverty. In Ethiopia 90 percent are MPI poor. The MPI captures deprivation directly in health and educational outcomes and key services, such as drinking water, sanitation and electricity. The multi-dimensional poverty index measured that the half of the world's poor live in South Asia [51 percent or 844 million people] and one quarter in Africa [28 percent or 458 million]. The multi-dimensional poverty index revealed that the countries with strong economic growth are having acute poverty. In India there are more MPI poor, than African continent. In eight Indian states, namely Bihar, Chhattisgarh, Jharkhand, Madhya Pradesh, Orissa, Rajasthan, UP and West Bengal having 421 million MPI poor and 26 poorest African countries 410 million MPI poor.

The global hunger index (GHI) report October 2010 of International Food Policy Research Institute (IFPRI) stated that the South Asian region has the highest regional hunger index at 22.9, ranks first with higher scores in indicating higher levels of hunger and malnutrition. The report indicates the low nutritional, educational and social status of women, which are the leading factors that contribute to a high prevalence of underweight children under five, one of the key indicators that form the basis of the ranking. The sub-Saharan African region ranks second with an index of 21.7. The 2010 global hunger index score fell by 14% in sub-Saharan Africa compared with 1990 score. The GHI report shows how the hunger situation has developed since 1990 at global, regional and national levels. Globally the GHI fell nearly one-fourth from 19.8 to 15.1. In spite of this positive trend the fight against hunger is not reaching its goals fast enough. The global hunger index report 2010 reveals that the first 1000 days of child's life—from conception to age two—are critical to tackling global hunger. According to global hunger index report 2010, the malnutrition among children under the age of two is one of the biggest challenges in reducing global hunger. It can cause lifelong harm to health, productivity and earning potential. The GHI report finds that 29 countries primarily in sub-Sahara Africa and South Asia have that level of hunger which is extremely alarming. Once a child passes the age of two, the negative effects of under nutrition are largely irreversible. The under nutrition among children has reached at terrible levels. About 195 million children under the age of five in the developing world are too small and under developed. About 129 million children under the age of five are under weight. The crisis of child under nutrition is a very serious dimension of dehumanization. The 2010 global hunger index report prepared by IFPRI, focuses on malnutrition among children under two to five years of age as a leading challenge to reducing global hunger and its adverse effects in the area of health, productivity and earning potential. The GHI report sets out the clear recommendation to inform and encourage the international community to take decisive action.

The millions of people in the poor and developing countries are illiterate, millions of children are out of school and millions of people are houseless. Near about one billion people lack improved and portable drinking water. Near about 2 billion people lack access to basic sanitation. The millions of children die due to malnutrition under the age of 5. Near about 60,000 women die every year during pregnancy or child birth period (report provided by UNICEF). Approximately out of more than 2.6 billion global workforces, more than 120 million are totally unemployed and 700 million are underemployed. The unemployment problem is increasing fastly in every region of the world (whether developed or underdeveloped countries). The slow pace of job creation even in those countries which are having relatively high growth rates left 500 million unemployed or underemployed in a region with a total labour force of 1.7 billion. The people of Africa, Latin America, and Asia (particularly in the under developing countries) are worst effected by inequality and poverty. The gap of economic inequality has been widened in spite of corporate capitalist economic reforms initiated in different countries of the world. The capitalist economic reforms proved anti nature and anti common people. This global human-inequality/dehumanization in ground exposing the hollowness of slogans of welfare of *aam aadmi*, [welfare of common man] social equality and social justice of democratic govts of the corporate capitalist world. The leading corporate players of the global corporate capitalist system claim that they have established the most just and equitable system. But the long historical experience and practice of global corporate capitalist system proved that this is the most unsustainable, irrational, inequitable and unjust social system of mankind, in spite of its great scientific-technological advances, growth and profit oriented global developments.

The corporate capitalist system upholds that mankind is the unique phenomena in the system of nature and nature is limitless. This approach has given license to corporate capitalists and their corporations (both national and multi-national) of the world to crudely exploit the natural resources and human resources in order to attain maximum growth. To projecting profit and capital maximization as the sole aim and purpose of human prosperity and progress, entitles the capitalists to take major share of wealth produced by the efforts of whole society. Thus creating a big gap between rich and poor in the world and turning the global common people into the victim of mass poverty, hunger, large scale unemployment, rising prices, ill health, illiteracy and global depression/recession.

The global corporate capitalist system pushed the world people by all means towards divisive lines and agendas in order to divide the mankind, which has caused civilization wars, national disputes, ethnic discrimination and clashes, racial discriminations and clashes, communal and religious clashes, tribal discriminations and clashes. The prevailing narrow/sectarian and fundamentalist human mindset, contradictions and clashes are the biggest stumbling block before human solidarity and sustainable advancement of mankind. The present global era demands that the divided house of mankind (divided into different nations, classes, groups, social formations, ethnic communities, social, religious, tribal and caste communities) should be united and integrated into a single human community on the basis of scientific-rational global humanism and nature-mankind friendly model/vision. The human deprivation, dehumanization, the global mass poverty, global mass hunger, global human-inequality and sectarian human/social contradictions are the outcome of global corporate capitalist system. The replacement of existing global corporate capitalist system is the need of the present global era. The present interdependent world need nature-mankind friendly global leaders who mobilize the whole mankind for transforming the present global corporate capitalist system, which is anti nature-anti mankind and the root cause of fundamental challenge of global climate change and global dehumanization.

3

Fundamental Challenge to Mankind: A Summary

The global corporate capitalist system is fundamentally unsustainable and contrary to the natural system of Planet Earth and majority of the mankind. The focus of this system is more growth and more profit by over using and over exploiting the natural and majority human part of the world by using pollutant technology, raw materials and energy. In fact, the global corporate capitalist system is in conflict/clash with natural system of Planet Earth and the fundamental creator of global warming/climate change and global dehumanization/ human inequality, the fundamental challenge of mankind. The challenge of global warming/climate change and global dehumanization/human cum social inequality clearly manifesting its most devastating and ugly face, even then the most developed nations and major developing nations are not serious about the gravity of this global crisis and refused to accept legally binding global climate treaty. The major corporate capitalist powers have failed to give a proper response to the fundamental challenge of global warming/ climate change and global dehumanization/human cum social-inequality. The corporate capitalist system is dealing with the challenge of global climate change with corporate capitalist perspective. The global climate change is not a market problem but it is an existential challenge. It cannot be solved by the market perspective and market means. The fight against the global climate change is not only a fight to save the mankind but a fight to save all forms of bio-life on Planet Earth. The threat of global climate change cannot be tackled effectively in the ongoing global corporate capitalist system which is based

92

on market mechanism, money capital, self interest, growth, centric and the responsible factor for the creation of global climate change and global dehumanization. No country can win alone the battle against global climate change and global dehumanization. The collective action is not an option but an imperative. Without the harmonious and balance development of nature and mankind the existing human or other bio-life on our Planet Earth cannot exist or sustain or flourish in long run. The nature and mankind have an equal importance. The world should immediately ban the anti nature-anti mankind unsustainable technology, energy, raw materials and fully promote and adopt nature-mankind friendly technology, energy and raw materials in order to build a safe and sustainable world in tune with natural world. This is vital because the natural world and human world are one — the mankind is the part of nature and cannot separate from it. So many activities on which the present world depends, abuse the natural system of Planet Earth. We no longer can afford to be so ambivalent. There is a historical need to replace pollutant energy system and raise energy efficiency worldwide. We should firmly and fully turn to clean and renewable sources of energy, such as wind, solar, geo-thermal, bio-mass, bloom, hydrogen and other new clean and renewable sources of energy and replace coal, oil and natural gas throughout the world. The present state of world energy is that the global economy and global corporate capitalist system are mainly based and depend on fossil fuels and traditional pollutant energy system. But the world is also in the early stages of two energy revolutions. First, shift to new energy efficient technologies. The larger energy saving potentials, including shifting from century old technologies, such as incandescent light bulbs and internal combustion engines to far more efficient technologies. The incandescent bulbs are being replaced by compact fluorescent lamp (CFL). But there are some reports that CFL is also to some extent harmful, which should be assessed scientifically. The second, energy revolution is the shift from economy powered by oil, coal and natural gas to power by wind, solar, geo-thermal and bloom, hydrogen and other energies are underway. The new energy economy is emerging and the traditional energy economy fuelled by oil, coal and natural gas is being replaced by clean and renewable energy such as wind, solar, and geo-thermal and bloom energy. Despite the global economic crisis, this energy transition is moving at such a fast pace and on a scale that we could not have imagined even few years back. It is a worldwide phenomenon, which needs more global focus and priority. The challenge of global warming/climate change forcing the mankind to tap clean and renewable energy sources of Planet Earth. The renewable energy sources are very vast. In US three states, North Dakota, Kansas and Texas have enough harness able wind energy to run the entire economy. In China the wind energy will likely become the dominant energy source. The Indonesia could one day get all its electric power from geo-thermal energy alone. Europe will be powered largely by wind farms in the North Sea and solar thermal power plants in North African deserts. In India the solar energy is becoming emerging source of energy in future under solar energy mission. In order to reduce 80% emission of green house gases by 2020, the first priority is to replace all coal and oil fired electricity generation with renewable energy sources. The 20th century was marked by the world energy economy as countries everywhere turned to oil. But the 21st century is turning to clean and renewable energy such as wind, solar, geo-thermal and bloom energy. The wind energy is the focus of energy economy and it is abundant, low cost and widely distributed. It can be developed quickly. The fossil fuel resources are pollutant and depleting but the wind resource of Planet Earth is clean and cannot be depleted. The worldwide survey of wind energy conducted by the Stamford University team of Christina Archer and Mark Jacobson concluded that the harnessing 1/5th of the earth's available wind energy would provide energy seven times as much electricity as the world currently uses. China has enough readily harness able wind energy to easily double its current electrical generating capacity. The US is also richly endowed with wind energy. Europe is already tapping its offshore wind for energy. The world wind energy generation is growing at a fast speed. The solar energy is also one of the world's fastest growing energy sources. The more and more countries are setting solar installation goals. The solar energy source is abundant. The International Energy Agency (IEA) the European solar thermal electricity association and Green Peace have outlined at the global level plan to develop 1.5 million megawatts of solar thermal power plant capacity by 2050. The pace of solar energy development is also accelerating. The heat in the upper six miles of the earth's crust contains fifty thousand (50,000) times, as much energy as found in all the world's oil and gas reserves combined. But only 10,500 megawatts of geo-thermal energy generating capacity have been harnessed worldwide. The US is contributing 50% to the world's existing geo-thermal energy capacity and the Philippines, Mexico, Indonesia, Italy, and Japan account for most of the rest geo-thermal energy generating capacity. The Iceland, the Philippines and El Salvador get 27%, 26% and 23% of their electricity from geo-thermal power plants. The other countries who are rich in geo-thermal energy are Chile, Peru, Colombia, Canada, Russia, China, Australia, Kenya, Ethiopia, Tanzania, Uganda and some places in India. The newly growing source of energy is plant based energy. As the oil and natural gas reserves are being depleting and harming the nature. The world's attention is also turning to plant based energy sources. The energy crops, forests by products, urban waste, plantation of fast growing trees, crop residues and urban trees and yard wastes, all of which can be used for energy generation and the production of automobile fuels. The restructuring of the renewable energy economy will be driven by the realization that the fate of human civilization and other bio-life depend not only doing so but on doing it at war level speed. We can expand the renewable energy use fast enough like the recent trends in the adoption of mobile phones and computers. In order to improve energy efficiency the whole world should develop and adopt more energy saving grid stations and advance lighting technology, like light emitting diode (LED). The energy saving measures that will offset nearly 30% growth in global energy demand projected by IEA. The building sector is responsible for a large share of world electricity consumption and raw material use. Worldwide building constructions accounts for 40% material use. The building construction on the basis of energy saving designs should be made mandatory worldwide. In new energy economy the buildings should rely on renewable energy for heating, cooling and lighting. To bring global change in transport sector, by introducing electric vehicles or highly fuel efficient vehicles and diversifying urban transport system. The transport sector must shift from gasoline-powered automobiles to plug-in gas-electric hybrids, all electric cars and light rail transit and high speed intercity rails.

The mankind will have to change the global corporate capitalist ways and means either by choice and deliberate design or through the force of circumstances. But the global historical experience of human society tells us that the fundamental social change only takes place when enough pressure has to be built up to that level which upset the existing balance of power. The time demands that the violation of natural or environmental laws and human rights violation be seriously treated as crime against humanity. The global indicators of environmental degradation and dehumanization are sounding the dangerous signal. Several prominent international scientists, environmental experts and social thinkers are giving serious warning for past 50 years about global warming/climate change and global dehumanization/ human cum social inequality. As per these warnings the situation is very alarming. The mankind reached at a cross road where it has only two choices or options. Either to stick to present global corporate capitalist system (which is the root cause and fundamental responsible factor for the creation of global warming/climate change and global dehumani-zation/human cum social inequality) and face the nature's fury, aggressive reaction of global people (especially the conscious people, poor, deprived and unemployed of the world) or ever biggest global disaster never seen in human history, which will perish/destroy the whole mankind and other bio-life on Planet Earth or save the nature and mankind from the global threat of global warming/climate change and global dehumanization/human cum social inequality. By immediately (without any delay, time is very short at our disposal, we are already late) stopping the emission of green house gases, protect environment/nature and rationalise the income difference in the whole world. The challenge of global warming/climate change and global dehumanization/human cum social inequality, facing mankind today is very dangerous and fatal. The mankind needs a complete unity among its different segments and social units in the world in order to counter such a dangerous and grave global challenge. The factor of global unity of mankind or human solidarity demands the permanent ending of all types of disputes/conflicts within the mankind throughout the world. The only way to achieve the human solidarity or unity of mankind and ending of all types of conflicts/disputes within the mankind lies in eliminating all sorts of existing injustice and discrimination in the world and do peaceful, democratic, rational and just compromising solution of existing global disputes/conflicts.

Keeping in view the dangerous challenge of global warming/climate change and global dehumanization/human cum social inequality, the nature-mankind friendly global movement demands from the industrialized developed corporate capitalist countries as under:

1. To change their extravagant and wasteful or luxurious lifestyle or way of life and also appeal to the developing countries to adopt the sustainable way of life.

2. To focus and invest maximum global capital to develop alternative renewable clean energy and clean technology. There is a historical need to fully change and revolutionize the global energy sector by forbidding its polluting energy technology and the adoption of green and clean energy technology.

3. To fundamentally change the global chemical based inorganic global agriculture system and to establish nature-mankind friendly global organic-cooperative [integrated multi-purpose people farming] agriculture system.

4. To constitute global fund for addressing the global hunger for more than one billion people.

5. To eliminate global dehumanization, global inequality, global poverty, global hunger, etc. by taking short-term measures and long-term fundamental systemic change of their creator the existing global corporate capitalist system.

The most industrialized developed countries (mainly responsible) in particular and developing countries in general are the contributors of global warming/climate change and global dehumanization/human cum social inequality. The developed and developing nations should take sincere and true united global action with genuine responsibility as per their share in the production of green house gases and degradation of natural environment/ecology of Planet Earth. Now it is not the time to carry out the approach of division between the developed and developing countries or the supremacy of national interests or protecting the global corporate capitalist economy. But it is the fundamental responsibility of mankind to save the Planet Earth and mankind or unique human civilization from the disaster of global warming/climate change and global dehumanization/ human cum social inequality and save from the dangers of outer space of the universe (from asteroids, comets, solar tsunami, solar radiation, etc.).

The governments of the different countries particularly the governments of rich developed nations should listen and understand the loud message of global scientific community and the voice of more than 700 billion global people, to act truly and sincerely as a part of human race or human species to save global environment (Planet Earth) and mankind. Otherwise be ready to face the music of global people in the form of global movement for saving nature and mankind. The rationality guides us that, it is better the creator of global existential threat to mankind and other bio-life, i.e. global corporate capitalist system should go and vacate global space for the "new dawn of new age", under nature-mankind friendly new global confederal system. The nature-mankind friendly global model/system will be the sustainable guarantee of the entire mankind, its children, and grand children. The nature-mankind friendly global confederal model humbly demands that the global people or global mankind should unite on the basis of save nature-mankind global agenda to counter democratically the global threat of global warming/climate change and global dehumanization/ human inequality and force the governments of the world (particularly the governments of industrialized developed rich countries) to fulfill the targets of Kyoto Protocol, Bali Action Plan and further proceed towards, to sign legally binding global treaties known as Save Nature-Mankind Global Treaties as under:

1. Globally cut the emission, by at least 50%, of green house gases immediately and rest within time bound period.
2. Global treaty of global peace and global no war pact.
3. Global treaty on global common security under single global confederal democratic system. Under this global treaty all the countries should dissolve their national defense systems and to uphold the global security system under global peace force, regulated by nature-mankind friendly global confederal system.
4. Reduce 75% global defense budgets and that amount of global defense budgets should be utilized for addressing the challenge of global warming/climate change and global dehumanization/human cum social inequality. The NATO should be dissolved and US should end its all military bases throughout the world including withdrawal of US led NATO forces from Iraq and Afghanistan.
5. Constitute nature-mankind friendly global sustainable development commission which will function under the nature-mankind friendly global confederal system. The repair work of environment damage and addressing the challenge of global warming/climate change and global dehumanization/human inequality should be done under this global commission.
6. Eliminate the global dehumanization/human inequality by ending irrational income distributions throughout the world.
7. Resolve peacefully and democratically all the disputes and conflicts between nations and intranational, existing in the world, on the basis of rational cum just compromise solution, such as J&K trilateral dispute, Maoist movement (Naxal uprising), Indo-Pak dispute, Indo-China border dispute, Sri Lankan-Sinhalese dispute, South Korean and North Korean dispute, Iraq and Afghanistan disputes, Middle-East dispute between Israel and Palestine and Arab countries, Russia and Georgia dispute, Russia and Chechnya dispute, US and Iran contention on nuclear issue, US-North Korean contention on nuclear issue.

8. Promote human solidarity and global unity of mankind and integrate/transform the divided mankind (in different communities and social groups/units) into a single global human community.
9. Religious-oriented global terrorist movements, militant movements, armed movements, unarmed ethnic or religious movements. To resolve peacefully and democratically all the global disputes in the light of nature-mankind friendly vision and rule out the military solution of political and social disputes of the world.
10. Forbid every polluting technology, energy and raw materials and develop nature-mankind friendly science and technology throughout the world and focus on global generation of nature-mankind friendly clean energy (renewable energy-solar energy, wind energy, hydel energy, geo-thermal energy, bio-mass energy, bloom energy, hydrogen energy, etc.) and promote energy efficiency throughout the world.

The market regulated global corporate capitalist model and state regulated socialist/communist model, both consider the mankind as the central phenomenon on our Planet Earth. This concept projects the people as the sole builder of history and human society. It disregards the equally important dimension of nature. In fact, without nature the mankind cannot survive and ignoring the human urge of equality is also one-sided approach without which man/mankind feel socially deprived. The corporate capitalist and socialist-communist model consider capital/money generation and wealth accumulation as their principal objective and manifestation of social development and progress. But the historical experience, facts and reality of human society shows that the nature and mankind are the two basic factors for creating everything in human society including capital or money or wealth. Without nature and mankind (which constitute the very life of human society) the progress and development of human society cannot move an inch ahead. The money or capital cannot do anything without nature and mankind. The capital or money is merely a medium of exchange not any supreme thing of human society. The mankind and other bio-life on our Planet Earth cannot exist, sustain, develop and flourish without the balanced development of nature and mankind. The nature and mankind have an equal importance without any distinction. Thus firmly observing the environmental laws (both natural laws and social laws for protecting the environment of Planet Earth) and the human rights is the historical responsibility of the whole mankind. The building of nature-mankind friendly and justice based global confederal democratic system is quite possible. Now time demands the replacement of existing unjust global corporate capitalist system and the establishment of nature-mankind friendly single global confederal democratic system in which all the nations will become integrated confederal part of the global confederal democratic system.

4   *Market Regulated Global Corporate Capitalist Model/System*

CAPITALIST SCHOOL OF THOUGHT

The present existential and fundamental two-fold global challenge of global warming/climate change and global dehumanization/human cum social inequality is the outcome of existing global corporate capitalist system (1750 onwards), which is based on capital (money), market regulation and self-oriented/self-centric approach. The corporate capitalist system (which is now reached at global corporate stage but still operating through multiple nation-states along with some international institutions and supra-national mechanism such as European Union or regional groupings) fundamentally developed on the basis of fundamental capitalist school of thought of world famous economist and philosopher Adam Smith. The market regulated capitalist theory of Adam Smith lays down three basic propositions in his world famous book "The Wealth of Nations" as under:

1. The capital or money is the basis of human society or human society is based on capital or money.
2. The market is the perspective or determining factor of capital
3. The man is selfish by nature. The historical experience and facts shows that these propositions or concepts or approaches are one-sided and wrong.

The first proposition that capital is the basis of human society or human society is based on capital (money) was fundamentally a wrong understanding and wrong approach. In fact, capital is only made up of money or currency. The mankind or human society historically always sustain or survive on sustainable nature and sustainable social conditions. This depends on the balance and harmony of nature and mankind. How can mankind or human society maintain and sustain its lives on money or currency made up of paper and coins.

The second proposition is that market is the perspective or determining factor of capital. The market is a corporate capitalist instrument to regulate the exchange of commodity production under the capitalist law of demand and supply (production and consumption). The market boosts the capital and capital owners and harms the interest of rest of the mankind in terms of money and wealth. This happens particularly during the capital market crisis or economic recession or depression. The market is not the determining or fundamental factor in the realm of sustainable development and progress but the nature and mankind are the equally fundamental factors of development, progress and human society. The third proposition or approach is that the man is selfish by nature. The experience or facts show that this proposition or concept or approach of Adam Smith is one-sided and wrong. In fact, man/mankind possesses two-sided basic nature, individual cum social. The individual aspect (biological) of man/mankind reflects the individual existence and individual physical living of man/mankind, while the social aspects reflects mankind's social living, social functioning (with the help of its own developed scientific-technological mechanism) and social organization. The study of anthropology (evolution and development of mankind), sociology and the historical developmental process of human society tells us that self interest is the basic instinct of animal kingdom and the human being since its evolution as *homo-sapiens* become bio-social (individual cum social) by nature. The bio-social nature of man/mankind is the only quality that distinguishes man/mankind from animal kingdom. When mankind become devoid of social aspect it turns out just like an animal in his/her lifestyle.

The capitalist projection of capital or money is the defining feature of social advancement that has highly and crudely boosted the human ambition of money making. It has become the main motivational factor of almost every human being (except few rational people). Today money has become the determining perspective, mindset, aim, priority and "means and norms". The universal acceptance of the rule of monetary growth as the sole standard of measuring social progress and prosperity is a serious defect of global corporate capitalist system.

The capitalist development model (now developed into a global corporate capitalist system) uphold the sole aim of capital/money/wealth maximization and for the attainment of this sole aim, the corporate capitalist countries and corporate capitalist class through their corporate companies (both national and multinational companies) always strive for the enhancement of growth-rate and growth-maximization. The global corporate capitalist system considers the growth-rate as the main criteria/rationale to judge the social development, progress and prosperity. This one-sided approach is a serious strategic loophole of global corporate capitalist system.

HISTORICAL EXPERIENCE OF MARKET REGULATED GLOBAL CORPORATE CAPITALIST SYSTEM

The market and corporate capitalist class regulated global corporate capitalist system fundamentally follows the principle of unsustainable growth maximization, money/wealth/capital maximization and profit maximization. The attainment of maximum growth, maximum capital/money/wealth and maximum profit under global corporate capitalist system mainly represents the interests of corporate capitalist class and maximization of self-interests. The historical experience of global corporate capitalist system tells us that the growth cum profit maximization and self-centric approach is contrary to the basic nature of mankind and fundamental factors (nature-mankind) of human society. The nature-mankind friendly social science guides us that the human society is based on nature and mankind. In order to attain the highest growth rate, the global corporate capitalist system crudely damaged the nature (natural environment of Planet Earth) and dehumanized the majority of mankind by creating the ever biggest human cum social inequality, global mass poverty and hunger (1 billion people are facing hunger in the world, 70% people in India are the victims of poverty and rising-prices, majority of African continent people particularly the people of sub-Saharan African region are facing severe hunger, malnutrition and absolute poverty and 1 billion people of the world are deprived from the pure drinking water). The global corporate capitalist system and its fundamental law of capital accumulation/concentration through profits monopolized the global capital/wealth/money/income in the hands of handful corporate capitalist class. The developed countries with hardly 15% of the world's population, today control over 85% of the world's material and financial resources, leaves less than 15% for 85% of the world people living in over 130 poor and developing countries.

The market regulated global corporate capitalist economic system having fundamental and potentially fatal weaknesses. It does not respect the sustainable yield and thresholds of natural system. It also favours the near term over the long term, showing little concern for future generation. The global corporate capitalist system is at present stealing the future of mankind and selling it in the present—glorifying and calling it gross domestic product or growth rate, which are its main rationale to judge the development and progress. The greed for money and wealth maximization led global corporate capitalist system to unsustainably misuse and overuse the nature (natural resources) and mankind (human resources). The approach of growth maximization and rapidly growing global population will further put more pressure on energy, food and water resources. The unsustainable global corporate capitalist model system created biggest imbalance between nature and mankind. The unsustainable global corporate capitalist development model brutally damaged the nature (natural environment of Planet Earth) and mankind. It has created biggest global human cum social inequality, dehumanization and mass impoverishment in the world. This resulted into the creation of two-fold fundamental challenge of global warming/climate change and global dehumanization/human cum social inequality. Today the mankind need a realistic view about the relationship between the economy and natural environment. The present world needs the nature-mankind friendly missionary and visionary leaders who could deeply see and grasp the complex, big and broader picture of nature and mankind. The historical experience of global corporate capitalist economy teaches us the great lessons that, now the present world should firmly adopt environment-mankind friendly or nature-mankind friendly economic model.

The global corporate capitalist system is also the creator of continuous global economic depressions (the panic of 1837, long depression of 1873–1896 and the well-known great depression of 1929–30) and recessions. The global recession would take a slowdown in global growth to three percent or less the four periods since 1985 qualify: 1990–1993, 1998, 2001–2002 and the global recession or financial crisis of 2007–2011, these financial crises badly affected the global economy and made millions of people unemployed throughout the world. The financial crisis of 2007–2011 had been called by leading economist the worst financial crisis since the one-related to the great depression of 1930. It contributed to the failure of key businesses, declines in consumer wealth estimated in the trillions of US dollars, substantial financial commitments incurred by governments and a significant decline in the economic activity. The global recession and financial crisis which has started in 2007 from US and expanded throughout the world, is still continuing in the world, particularly in US, Europe (Greece, Ireland, Italy, Portugal, etc.) and some other parts of the world. The reduced economic growth due to global financial crisis has halted pace of globalization, increasing protectionist pressure and financial fragmentation. The global financial crisis has accelerated the global economic rebalancing. In the west the biggest change is the increasing role of state power in economy. The global financial crisis demands to better regulate the world economy. Now the financial landscape has become global and multipolar— US and European Union in the West, Russia and GCC nations in central Asia and Middle East, China and India in the east. The recent global financial crisis heightens interest in less leveraged finance. Such a global and multipolar financial trend signals a relative decline for US power and likely increase in market competition and complexity. These multiple financial centres may create redundance that helps to insulate market against financial shocks and currency crisis.

The corporate capitalist countries are trying very hard to overcome this financial crisis and global economic recession but it is very difficult to solve the global economic crisis permanently because it is a regular trend and feature of global corporate capitalist system which will occur from time to time as happened in the past, up to the death or total transformation of global corporate capitalism.

The long historical process of capitalism shows that the corporate capitalist system (a highest stage/phase of capitalism) which began with the industrial revolution and passed through the various phases of development: (1) industrial phase (handicraft, steam engine, electric energy), (2) industrial competitive phase, (3) colonial phase, (4) post colonial phase, and (5) global corporate capitalist phase, which is now going on throughout the world. Today the mankind is living within a global corporate capitalist system which is operating through multiple corporate organizations of nation-states. The entire nation-states of the world follow the same corporate principles in their thinking, functioning and organization. The only difference among them pertains to their respective variations in developmental level, territorial and population sizes. At social level they follow the same ideology, politics, economics, culture, diplomacy and security system along with the difference of political mechanism and regulation, methodological difference, lingual and community (both national and pre-national communities) difference, ambition of becoming super power, some historical differences inherited from the evolutionary process of mankind or human society/ human civilization. There is no human activity, lifestyle or way of life independent from the framework of global corporate capitalist system. The post-1945 scientific-technological revolution, particularly the revolution in information technology, internet technology and communication techno-logy have globalized the human interaction, information, knowledge and moment. The age of globalization is conti-nuously transforming the nation-states and threatening their usefulness and viability. Mr Keylor commented that the defining characteristic of this new era is the growing power and importance of a diverse set of non-governmental global organisations that transcended national boundaries and escaped the supervision of the national political entities that had dominated the history of global relations for the past four hundred years with the full sovereignty of nation-states. The globalization of capital and finance means that the national governments no longer have economic control of their nation with the focus on export and trade as the key drivers of economic activity issues, such as the international stock and trading markets, currency values and global political contexts are greater determinants of economic conditions within a country than internal factors. In corporate capitalist system the economic globalization is asserting a growing influence over employment patterns within nations. Traditionally one of the most important areas of the government control and regulation was employment availability. But now the conditions are increasingly determined by the global factors. The powerful global corporate capitalist companies pursue comparative advantage in respect to employment. These global corporate companies are always looking for high productivity and low costs. The national governments cannot maintain control over their labour/employment sector. There is pressure for governments to open up their markets under a free trade global system. This means that the local companies must compete with imported products that are often cheaper and resulting in loss of local jobs and business. The borderless world of electronic communications means that the national governments can less readily control the information that is available within any nation. The speed of information through telecommunication, media (both print and electronic), radio and air transport brings worldwide participation in events and issues. The internet is a global system which is providing to whole mankind the oceans of knowledge and information on every subject from every nook and corner of the globe. The efforts to politically control the internet system have achieved only partial success.

The large sums of finance capital, such as shares, futures and currency speculations can be shifted rapidly out of a region. The great flow of mankind around the world has put pressure on the capacity of national governments to control their borders and populations. The refugees of manykind and circumstances, the political asylum seeker, war refugees, climate refugees and economic refugees to find a place where they can settle new lives. Under globalization certain areas that have been always the responsibility of the nation-states, now are being privatised and taken over by the transnational/multinational private companies. The contemporary global era is characterised by the economic expansion of global corporate capitalist companies. The generation of globally competitive companies are emerging from the new powers, solidify their position in the global market; Brazil in agri-business and offshore energy exploration, Russia in energy and metals, India in IT sector, pharmaceuticals and auto parts and China in steel, home appliances and telecommunication equipment. The top 100 new global corporate companies listed in 2006 and among them 84 companies established their head quarters in China, India, Russia and Brazil. The transnational or multinational corporations are understood as being corporations that have originated and are largely based on one nation with subsidiaries and operations spread across many countries and regions. It is estimated that there are more than 10,000 multinational corporations around the world. The top 15 multinational corporations in terms of world revenue ranking are as under: (1) Wal-Mart, (2) Exxon, (3) General Motors, (4) British Petroleum, (5) Ford Motor, (6) Enron, (7) Daimler Chrysler, (8) Royal Dutch/Shell, (9) General Electric, (10) Toyota Motor, (11) Citi Group, (12) Mitsubishi, (13) Mitsui, (14) Chevron Texaco, (15) Total Fina Elf. The combined sales of the world's top 200 multinational corporations are far greater than a quarter of the world's economic activity. The resource companies cover mining, oil, energy, timber and other resource extraction and processing companies. The largest of these companies are completely global, extracting resources in different regions and selling to worldwide markets. Due to the type of products extracting these companies, frequently operate in remote, less populated and underdeveloped regions, such as deserts, forests, oceans and Islands where they use high technology means of extracting the raw resources of Planet Earth. They usually are shipping the raw resources to other locations for processing, then using distribution chain to sell the final product. These global corporations on the one side contributing to the world through providing needed products, such as fuel and metals, as well as jobs and incomes but on the other side these companies create many global problems through pollution and depletion of the Planet Earth's limited resources.

They sometime interfere with complete eco-system, such as logging in forests that are the homes of people, rare species of animal and important plants. In other words these global companies pose a threat to earth's bio-diversity. The products and bio-products of these companies on the one hand give benefits to the world, mainly to them and on the other hand become poisons, causing pollution, emission of green house gases and threat of global warming/climate change. The transnational industrial companies comprise of those corporations that make goods in factories in various locations and then sell them around the world. The activities of global industrial companies include a wide range of activities from steel production to domestic consumer goods. Their activities are one of the main forces in the globalization system, having effects on employment pattern, application of new technology and consumer culture. Previously the raw resources were all brought into one place for the factory process. This resulted in the growth of great industrial centres of production in the national stage of competitive capitalism. But now in the global corporate capitalist stage the manufactured products are increasingly made from component parts produced in specialist factories in widely dispersed countries and then brought together and assembled for sale in other countries. The globalization has intensified this trend towards a differential specialist production model. The combination of new economics, sophisticated machinery and the capacity for rapid communication between distant centres have all facilitated the successful spread of this globalised production model. The one important change in production has been the trend toward reorienting successful national products toward global market, which in turn change the identity of the products. The financial companies are also one of the key drivers of economic globalization as they expand their activities around the world. They include the banks and finance credit companies and investment and insurance corporations. The transnational service industries are the most recent and most rapidly expanding sector of the global economy. The unprecedented global economic growth under global corporate capitalist system will continue to put pressure on a number of highly strategic resources including energy, food and water. The world bank estimates that the demand for food in the world will rise by 50% by 2030 as a result of growing world population, rising affluence and the shift to western dietary preferences by a larger middle class.

The global corporate capitalist system promoted self interest as the motive force in the human society. The self-centric approach created the motivation of selfishness throughout the world. Thus increased the trend and tendency of personal greed and gain and neglect the other important human dimensions of social or common interests, urge of equality, saving of nature and mankind, etc. The nature and mankind are the two most important factors of human society, which constitute the social capital. Without nature and mankind nothing has any value. The prevailing capital or money is only a medium of exchange, operating in the form of currencies and commodities (commodity means product or production meant for market sale). Today in the global corporate capitalist stage the capital exists in the form of big concentration of money, large accumulation of commodities and stock market (capital market) shares.

The change and development in the human society takes place through two-sided interaction between nature and mankind and within the mankind (between different social units, classes, groups, communities, national entities, etc.) through the process of unity and struggle alternatively. Sometimes struggle plays primary role and unity remains in secondary position and sometimes unity plays primary role and struggle remains secondary. This law rule out the permanent primary position of unity or struggle. The same law is also applicable in the domain of nature (this law was propounded by late R.P. Saraf, a great social thinker of nature-human centric people movement). This two-sided interaction shows that the human resources and natural resources constitute the highest valuable thing or the supreme phenomenon of the human society, deserve the status of capital. The definition of capital is that, it is the supreme and the highest valuable thing in the human society. The corporate theory of capital, as the accumulation of money and wealth is only a false and one-sided theory because money constitutes only a medium of exchange of global corporate capitalist system.

The corporate capitalist system adopted its basic economic principle of profit optimization as the only determinant of social interest and concerns. The sole objective of profit optimization combined with the approach of self-interest has set in motion the race for money making all over the world through corruption, black money or legal means. The role of corporate sector in the money laundering business can be seen from the recent cases of corporate frauds and scandals in the world's topmost multinational companies. In India biggest fraud done by the Satyam company, the criminal and callous attitude of chemical company of US, Union Carbide and now takeover by US company Dow chemical company in Bhopal, Madhya Pradesh, India.

The global corporate capitalist system stands for developing the capital and capital owners and upholds the corporate capital management and control. The global corporate capitalist system practises the authoritarian style, methods and super power designs. Now the world is divided politically, economically, culturally between global super power, i.e. US and regional super powers, i.e. China, Japan, Britain, France, Germany, Russia, India and poor developing nations. In the global corporate capitalist system the rule of "might is right" is prevailing throughout the world. This can be seen from the 40 years experience of cold war of super powers (US and USSR). After the collapse of socialist block and disintegration of Soviet Union, the US remained sole global superpower, which has tried to make unipolar world but the hard reality of present world is that it is an interdependent multi-polar world which is heading towards a global system. The US being a sole super power of the world (particularly in the period of both the Bush senior and junior, leaders of US Republican Party) has damaged the cause of "global peace", "global human solidarity" and created very negative results in the world by adopting the unilateral and belligerent approach. Now the period of super powers is over. A global corporate capitalist multi-polar world has emerged with the rise of China, India, and others. Ultimately the nation-states will no longer exist. The power will be more dispersed with the newer players, bringing new rules of the game. The traditional western alliances will weaken. More countries may be attracted to China's alternative development model rather than emulating western model. China is poised to have more impact on the world over the next 20 years than any other country (if the global mass revolution for the replacement of global corporate capitalist system will not happen, which is the historical need of present global era). In fact, the Chinese model is an integrated part of global corporate capitalism with some variations to US and western model but it will definitely fail and collapse with global corporate capitalist system as it is also anti nature-anti mankind. China is a biggest creator of pollution/green house gases and dehumanization/human cum social inequality along with US, most developed and big developing countries (India, Brazil, South Africa, etc.). If the current trends persist by 2025 then China will have the world's second largest economy and will be a leading military power. China could also be the largest importer of natural resources and the biggest destroyer of natural environment. India probably will continue to enjoy the rapid economic growth and will strive for a multi-polar world in which India is one of the poles. China and India have to decide the extent to which they are willing and capable of playing the increasing global roles. China and India for the first time since the 18th century are set to be the largest contributors to global economic growth. These two countries are likely to surpass the GDP of all other economies except US by 2025 but they will continue to lag in per capita income for decades. The growth projections for China, India, Russia, and Brazil have collectively matched the original G-7's share of global GDP by 2040–2050. The eight largest economies in 2025 will be in descending order: US, China, India, Japan, Germany, UK, France, and Russia. But China has emerged as a new financial heavyweight, claiming 2 trillion dollar in foreign exchange reserves in 2008. Russia has the potential to be richer, more powerful and more self assured if it invests in the development of human capital, expand, diversifies and integrate its economy with global economy. Russia could experience a significant decline if it fails to take these steps and the prices of oil and natural gas remain in the range of 50-70 dollar per barrel. No other countries are projected to rise to the level of China, India, Russia and none is likely to match their global clout. But the political and economic power of

Brazil, South Africa, Indonesia, and Iran can increase. The rapidly developing countries have created Sovereign Wealth Fund (SWF) with the aim of using their hundreds of billions of dollars worth of assets to achieve higher returns to help them in economic crisis. Some of these funds returned to the west in the form of investments in order to promote greater productivity and economic competitiveness. However, the foreign direct investment by these emerging economic powers in the developing world is increasing significantly. The sovereign wealth fund constitutes the capital which is generated from govt surpluses and invested in private market abroad. Since 2005 the number of countries with SWF has grown from three to over forty and the aggregate sum under their control from around 700 billion dollars to 3 trillion dollars. The range of functions served by SWF also expanded as many of the nation-states that created them recently, have done so out of a desire to perpetuate current account surpluses or to cultivate inter generational savings rather than to buffer commodity market volatility. If the current trends continue the SWF will swell to over 6.5 trillion dollars within years and 12 to 15 trillion dollars within decades exceeding total fiscal reserves and comprising some 20% of all global capitalization.

Today wealth is moving not just from west to east but is concentrating more under state control. In the wake of 2008 global financial crisis the role of state in the economy is gaining more appeal throughout the world. The nation-states that are beneficiaries of the massive shift of wealth are China, Russia, and Gulf states. These nations are not following the western liberal model for self development but are using state dominant corporate capitalist mix model. The state dominant corporate capitalist mix model is a system of economic management that gives a prominent role to the state. In the western economies the enhancement of state role is underway as a result of current global financial crisis. It may reinforce the emerging countries preference for greater state control and distrust of unregulated market system. The governments that highly manage their economies often have an interest in industrial policy. The China, Russia and Gulf states have state plans to diversify their economies and climb the value-added ladder into high technology and service sectors. In the early 1990 many economists predicted that state-owned enterprises would be a relic of 20th century. But they proved wrong. The state owned enterprises are far from extinction and in many cases seek to expand beyond their own borders, particularly in commodity and energy sectors. The increasing role of the state in emerging markets has contrasted recently with nearly opposite trends in the west, where the state has struggled to keep pace with private financial engineering, such as derivatives and credit swaps. The financial engineering has injected an unprecedented degree of risk and volatility into global markets. The greater controls and global regulation— a possible outcome of the current global financial crisis— could change this trajectory. Although a gap on the role of the state in the economy is likely to remain between the west and rapidly emerging economic powers. The emerging economies despite boom are refusing to allow currency appreciation, ultimately albeit unsustainable cycle of imbalances. The nation-states entering private markets are doing for the prospects of higher returns. The Gulf Co-operation Council (GCC) and Chinese govt amassed huge assets. They have increasingly used various forms of sovereign investments.

China, India and Russia have not fully adopted the western liberal model. But they are using both the institutions—state and market for their development. South Korea, Taiwan and Singapore also used both the state and market institutions of global corporate capitalist system to develop their economies. However, the impact of China and Russia by adopting this path is potentially much greater because of their sizes and approach. The South Asian region and sub-Saharan African region still remain most backward, vulnerable to economic disruption, population stresses, civil conflicts and politically instable regions of the world. Europe and Japan will continue to far out distance from the emerging power of China and India in per capita income but they have to do hard struggle to maintain robust growth rates because of the size of their working age populations will decrease. The number of countries with youthful age structure in the current arc of instability is projected to decline as much as 40%. Three of every four youth-bulge countries that remain, will be located in sub-Saharan Africa, in the core of Middle East, scattered in southern and central Asia, and in Pacific islands.

The national governments of global corporate capitalist system follow the rule of "might is right" under the banner of democracy but in real sense the democracies are regulated by few top politicians of major political parties, top bureaucrats, corporate capitalists, rich classes, criminals and mafias, black money holders, unprincipled and characterless politicians, casteist and communal forces and somewhere in the world the military rules are prevailing. But the common masses and activists of different political parties in the world are only meant for casting votes in their favour and increase the strength of their rallies. The common people or party activists don't have any power to select candidates for elections and party office bearers (may be there is some improvement in the advance democracies). In the name of high command, they established the crude form of undemocratic and authoritarian systems with the sign board of democracy. But in reality all the corporate capitalist democratic systems of the world strangulated the inner and outer democracy which required in the present global era. The honest and pro- people dedicated leaders have been isolated and sidelined by the powerful lobbies but the favourable time for honest and efficient leaders will come very soon. The honest, rational, nature-mankind friendly and far-sighted leadership is the historical need of the present global era. Such leadership will definitely emerge with the support of highly democratic conscious global masses. What is the future of liberal democracy in present world? Subjectively it does not look too bleak. There are no powerful movements challenging the corporate capitalist liberal democracy in principle except Islamic theocracy. The second half of the 20th century was the age of military dictatorship which was a far greater danger to western and independent former colonial regimes. The 21st century does not look quite so favourable to them. The communist countries primarily adopted the political model of one party dictatorship but after the long negative experience of one party dictatorship and great mass resentment and mass uprising the one party dictatorship in almost all communist countries (except China, North Korea, and

Vietnam) has been overthrown by the people. In these communist countries the liberal democracy and market system established. But in fact, the real nature of corporate capitalist liberal democracy and market system has been exposed and proved anti nature-anti mankind. The global corporate capitalist system and its so-called democracy has become outdated and heading towards failure and collapse.

The principles of "might is right" and "whatever means but the end should be justified" (i.e. non conformity of means and ends) develop the mindset of power seeking at all cost by foul or fair means, finally led to the criminalization of politics, communalization of politics, role of money or black money in politics, business of mandate distribution by the political parties, ultimately corrupted the politics and economy of the world.

This trend and practice degenerated the whole political, economical, security, diplomatic and cultural processes and made the whole global corporate capitalist system and its governments corrupt, hegemonic and dominate to weak. The growth of few big and powerful rich people, dehumanization of majority mankind, deprivation, sectarian division and suppression of the poor and common people is the order of today.

The latest report of Transparency International (2009) exposed the corruption level of the different countries. The representatives of corporate capitalist class, black money holders, corrupt leaders, mafias and criminals are entering in the national parliament or national assemblies or national congress and in the state or federal units of the world. In the pro-corporate capitalist national governments, the most of the political parties and political leaders (except some honest leaders) in every country are mainly concerned about amassing wealth and holding of everlasting power by fair or foul means and least concerned to the poor masses and mankind's general interest. This is proved in the global food conference held at Rome, Italy 2009, in which no big and rich country participated. The indifferent attitude towards the nature and mankind of Planet Earth is proved by the greatest fundamental challenge of global warming/climate change and global dehumanization/human cum social-inequality. Still the most industrialized rich nation who are mainly responsible for the emission of the green house gases and global warming/climate change and global dehumanization/human cum social inequality have no first priority to save nature-mankind from the global warming/climate change and global dehumanization/human cum social inequality, a global threat facing entire mankind. They along with corporate capitalist class mainly concerned about money and wealth maximization and power monopolization. They are continuously increasing the wasteful and unproductive expenditure on militarization and manufacturing of nuclear and mass destruction weapons. Those who have huge stock of nuclear weapons are advocating nuclear non-proliferation under NPT and CTBT, both are the discriminatory treaties.

The leaders of global corporate capitalist system always speak about poor and common people but by all means mainly serve the capitalist, rich and dominant classes. Sometimes they announce popular policies and economic packages and immediate reliefs in order to survive among the people but never ready to dismantle or fundamentally restructure or remodel the corporate capitalist irrational and unjust system and its outdated anti-nature and anti-mankind global corporate capitalist framework which is the root cause of the global warming/climate change, dehumanization, human inequality, poverty, hunger, unemployment, rising prices (due to capitalist market phenomenon), malnutrition, etc. The holding of large scale black money (which is the outcome of global corporate capitalist system) by many top politicians, top capitalists, top bureaucrats and top mafias of the different countries severely affected the global economy. The huge black money (trillions of US dollars) has deposited in the Swiss Banks, Banks of Belgium, Banks of Luxemburg, etc. (banks operating in Switzerland, Belgium, Luxemburg, etc.). Neither the UN, nor the world powers nor the national govts are seriously forcing these countries to expose and return the black money of the world, deposited in their banks. The UN has adopted the convention against global corruption regarding the information and returning of black money but no body is serious to get back black money deposited in the foreign banks and illegally invested in different areas and sectors of the world. The whole black money of the world deposited in the banks of abovesaid countries and illegally invested in the world should be confiscated under the UN supervision. The confiscated black money should be used for the repairing of damaged natural environment of Planet Earth, global initiative to stop global warming/climate change and to invest for the elimination/ eradication of global hunger, poverty, unemployment and dehumanization. The nuclear and mass destruction weapons should be dismantled under the UN supervision. The US should end its all military bases in the world and dissolve NATO immediately. All the countries should immediately reduce their 50% military budgets under save nature-mankind global peace and no war treaty in first phase and the rest 25% budgets in second phase. Through these great positive actions in favour of saving nature-mankind, we can save or generate trillions of money for countering the global threat of global warming/ climate change, global dehumanization, human-inequality, hunger, poverty, unemployment, and malnutrition. This will be a great service for saving nature and mankind of Planet Earth.

Today the fundamental social reality of corporate capitalist world is that the post-1945 scientific-technological revolution has changed the world of sovereign nations into an inter-dependent corporate capitalist world, which is operating through multiple nation-states. The scientific-technological revolution in key technologies (space technology, informational technology, bio-technology, energy technology and material technology) has transformed the sovereign nation's scattered world into an interdependent globalized world. The techno-logical and productive transformations are quite obvious. Think only the speed of the communication revolution, which has virtually abolished time and distance. The internet is only few years old but it brought a great information and knowledge revolution, which is expanding and spreading the old as well as latest information and knowledge in the whole world in a supersonic speed and continuously bringing closer the scattered mankind into a global network. The unprecedented transformations of the past half century or descend for a closer look at the factors affecting the trend of war, peace and the balance of global power at the outset of the 21st century. The general trends are not necessarily guides to practical realities. It is evident that in the course of the 20th century the world people ceased to be overwhelmingly ruled, as it were from the top to bottom by hereditary princess or the agents of foreign powers. It is now come to live in independent nation-states whose governments claimed legitimacy by reference to the people or the nation (including the totalitarian regimes) in most of the cases claiming confirmation through real or bogus elections or periodic mass public ceremonies that symbolised bond between authority/regimes and people. Now the people have changed from being subjects to being citizens.

The social aspects that are relevant for global future are in dramatic decline and fall of the peasantry which formed the great bulk of the mankind as well as the foundation of economy until the 19th century. The corresponding rise of a predominantly urban sector especially the hyper cities. In 1900 only 16% of the world population lived in towns, in 1950 that had risen to just 26%, today it is 48% in the developed countries and many other parts of the globe. It is true that the old developed countries are heavily urbanised but they are no longer typical of current urbanisation, which takes the form of a desperate flight from the countryside into hyper cities. Today only ten of the world's largest fifty cities and only two of the eighteen world cities whose population stands at ten million or more in Europe and North America. The fastest growing cities over one million are in Asia (20), Africa (6), and Latin America (5). This process dramatically changed the geographic, political, economic and cultural balance between highly concentrated urban and geographically spread out rural population. The world of oral communication has been replaced by a world of universal reading and writing by hand or machine.

The globalization of the corporate capitalist system has become the new reality of the present world. The rapidly globalising the world economy based on transnational private firms that are doing their best to live outside the range of nation-states law and taxes, which severely limits the ability of even big governments to control their national economies. Under the market regulated global corporate capitalist system the nation-states are actually abandoning many of their most traditional direct activities to profit making private contractors. There is decline in acceptance of the nation-state legitimacy and the voluntary acceptance of obligation to the ruling regimes and their laws by those who live in their territories, whether as citizen or as subject. The obedience of citizens is eroding quickly. Whether any state today could engage in major war with conscript armies ready to fight and die for their country to the better end is doubtful. The few developed western nation-states or better governed pro nature-pro people more democratic nation-states can for some time rely on law abiding and orderly people (except criminals and mafias). In fact, the extraordinary rise of the technological and other means of keeping the people under surveillance at all times (through closed circuit TV cameras, phone tapping and access to personal data and computers) has not made the nation-state more effective, though it has made citizen less free. This has been taking place in an era of dramatically accelerated globalization, which is giving rise to regional disparities around the world. The corporate capitalist globalization produces unbalanced and asymmetric growth. The mankind is facing the problems of the 21st century with the collection of political mechanisms dramatically ill-suited to dealing with them. They are effectively confined within the borders of nation-states, whose numbers are growing and confront objectively an inter-dependent global world which lies beyond their range of operation. It is not even clear how far they can apply within a vast heterogeneous territory which possesses a common political framework, like the European Union. They face and compete the world economy effectively operating through different units (transnational firms) to which consideration of political legitimacy and common interests do not apply and bypass the national politics. They face the fundamental problem of the future of the world in an age when the impact of human action on nature and the globe has become a force of geological proportions. The disintegration of the USSR left the US as the only superpower, which no other power could or wanted to challenge. But the hard fact is that the present world is too complicated and it is very difficult for any single nation-state to dominate fully the world. In fact, the US economy forms a diminishing share in the global economy and it is vulnerable in short term as well as in long term. The OPEC decided to put all its bills in euros instead of in dollars. This indicates the great decline of US corporate capitalist economy. It also underline the contradiction between those aspects of contemporary life that are subject to globalization and the pressure of global standardisation [in science and technology, economy, technical infrastructures and cultural institutions] and those which are not the state and politics. The globalization logically leads to an increased flow of labour migration from poorer to richer regions but this produces political and social tensions in many of the nation-states affected, mostly the rich countries of the north Atlantic region. Unlike the movement of capital, commodities and communications, the nation-states put effective obstacles in the way of global capital and commodities and human resource movement. The unprecedented shift of wealth and economic power from west to east is under way. This shift derives from two sources. First, the increase in oil and commodity prices, which has generated wind fall profits for the Gulf States and Russia. Second, the low cost labour has shifted the focus of manufacturing and service industry to Asia. The shrinking economic and military capabilities may force the US into a difficult set of trade off between domestic and foreign policy priorities. The domination of US superpower will be reduced in coming years. The most striking global change created by the economic and developmental globalization is the growing shift of the centre of gravity of the global economy from the US and Europe to Asia. In terms of size, speed and directional flow, the global shift in relative wealth and economic power now under way roughly from west to east. This is still in its early stages but this process is accelerating. The growth of global economy over the past thirteen years has been pulled along largely by the Asian dynamos, notably by the extraordinary rate of growth of industrial production in China with 30% rise in 2003 compared with 3% for the world and less than 0.5% in North America and Germany. Australia, France, Italy and UK had negative growth [CIA world fact book up to 19 Oct 2004]. It is very much clear that this has not yet greatly changed the relative weight of Asia and the North Atlantic—the US, the European Union and Japan, between them continue to account 70% of the global

GDP—but the size of Asia is already making itself felt in terms of purchasing power. The South Asia, South-East Asia and East-Asia already represent a market about two-thirds larger than the US economy. This is naturally a central question to the global prospects of the 21st century. The prospect of global peace in 21st century is brighter than the 20th century with its unparalleled records of world wars, wars between nations and other theatres of deaths on large scale. The typical 20th century form of warfare that waged between nation-states has been declining sharply. At the movement no such traditional wars between nations is taking place. But the internal conflicts are going on in various areas of the Africa and Asia, where the internal stability and cohesion of the existing nation-states are at risk. In fact, the possibility and danger of a major global war probably arising out of the unwillingness of the US to accept the emergence of China as a rival superpower has not receded, although it is not immediate possibility. Indeed the chance of avoiding such a global conflict or better than the chance of avoiding Second World War. The actual danger and challenge to global stability or to any stable nation-state from the activities of the pan Islamic terrorist networks—against which the US and its allies proclaimed their global war on terror—is negligible. Though the terrorist groups kill much larger number of people but the risk to life they present is statistically minimal than from the challenge of global climate change and global dehumanization. Unless the terrorist groups gain access to nuclear weapons, which is not unthinkable but not an immediate prospect. The problem and threat of global terrorism call for cool heads to tackle it. We should not tackle the global terrorism with cruelty and hysteria but the world should tackle it with rational approach by focusing on its root cause.

Today there are strong possibilities of global disorder, global disaster and human calamities due to the threat of global climate change and global dehumanization, which is the outcome of unsustainable anti nature-anti mankind global corporate capitalist system. The nation-states and market regulated global corporate capitalist system is continuously becoming ineffective to tackle the global challenge and global objective reality of interdependent world. Today the problem is more difficult because the rapidly growing inequalities created by market regulated global corporate capitalist system has created great imbalance in the world. It has been recently observed that not even the most advanced military establishments could be expected to cope with a general break down of legal and social order and the crisis of nation-states. In long run (after one or two decades approximately) the nation-states will no longer exist or will exist in a fully diluted forms. The new global reality and fundamental global challenge of global climate change and global dehumanization will force to dilute and dismantle the monopoly of global corporate capitalist and pro-corporate capitalist political players and centralized authoritarian polity over global political power and global corporate capitalist economy. The global power and global corporate capitalist economy will be fully democratized and transferred to global people through global mass revolution.

There is no longer a global power system in a position to keep at bay a general collapse into global war. The disintegration of USSR and military superiority of US have ended the global block power system. The US and its supporters claimed that this is the opening of new era of global peace and economic growth under the US empire. They compare wrongly the pox Britannica 19th century British Empire. Historically the empires have not created peace and stability in the world. The empires have always justifie themselves, claiming to spread civilization or religion or freedom to the victims of oppression. Whether the aim of global domination by a single nation-state is possible and whether the military superiority of the US is enough to establish and maintain it. In fact, this is not possible in a globalized interdependent world. The arms have often established empires but it has taken more than arms to maintain them. The most historic empires ruled indirectly through native elites. But when they lost their capacity to win enough friends and collaborators among their subjects, the arms are not enough. The second half of the US hegemony in 20th century rested not on military might but on the enormous wealth and central role of its giant economy expanded in the world. The military strength underlines the economic vulnerability of US, whose enormous trade deficit is maintained by Asian investors, whose economic interests in supporting a falling dollar is rapidly diminishing. It also underlines the relative economic clout of the European Union, Japan, East Asia and even the organised block of third world. In the World Trade Organization the US can no longer negotiate with clients. Indeed, may not the very aggressive, indicate a basic sense of insecurity about the global future of the US. The old British empire is not and cannot be a model for the US aim of global supremacy. When the age of western Empires ended in the mid-20th century the Britain recognised the wind of change. Will the US learn this lesson or will it be tempted to maintain an eroding global position by relying on politico-military force? In doing so, it will not promote global order but disorder, not global peace but conflicts, not advance the civilization but of barbarism. The armed power of the nation-states has proved incapable to maintain control of their territory. There is indeed a general crisis of nation-state power and state legitimacy even on the home territories of old and stable European nation-states such as Spain and UK. There is no prospect of a return to the imperial world of the past. The age of Empires is over and dead. The people of the present world have to find another way of organising the globalized inter-dependent world of the twenty-first century. The organising of globalized inter-dependent world demands that to assess the present state of affairs of nation-states. After the end of cold war, dissolution of socialist block and disintegration of USSR, the unilateral attempts have been taken to establish new global order under US domination but have not succeeded. The year 1990 witnessed notable Balkanisation of large parts of the old world, mainly through the disintegration of USSR and the communist regimes in the Balkans. The largest increase in the number of globally recognised sovereign states. This period has also seen the rise of the failed states. The collapse of effective government control or a situation of endemic internal armed conflicts in some regions of several nominally independent nation-states in Africa and former communist nation-states and also in one region of Latin America. Nevertheless the largest regions of the world remain unstable. The important element affecting the nation-states is the extraordinary acceleration of the process of globalization in recent decades and its effect on the movement and mobility of mankind. Ideologically the West undoubtedly benefited as the big advocate of freedom and democracy against tyranny except in those regions where it was allied with the dictatorial countries, which lacks freedom and democracy. But the western signboard of democracy and freedom now fully faded and exposed by its belligerent military action against Iraq, Afghanistan and Libya.

Today the globalization is forcing the world to adopt supranational model or integration of nations, international and regional economic cooperations, international and regional formations and groups, e.g. (1) The European Union (a supranational corporate capitalist model), (2) The G8 group (the economic group of most developed capitalist countries), (3) The G20 group (the group of developed and major developing countries, (4) The World Trade Organization (WTO), (5) Organization of Economic Cooperation and development (OECD, a group of 30 countries), (6) The Asia-Pacific Economic Co-operation (APEC), (7) The Association of South-East Asian Nations (ASEAN), (8) The South Asian Association of Regional Co-operation (SAARC), (9) The Organisation of African Unity (OAU), (10) The Organization of Islamic Countries (OIC), (11) The Organization of Petroleum Exporting Countries (OPEC), (12) The merger of east and west Germany, (13) The east African community, and (14) G77 group.

The above formations and process indicate the globalization of corporate capitalist system. The global corporate capitalist system may develop into a super global capitalist system but it does not revert to a pro-nature and pro-poor or common people model. This is because the global corporate capitalist system has its own laws of development and regulation, i.e. the law of capital/money/profit/wealth maximization or the law of capital/money accumulation and its concentration around the capitalists. The corporate capitalist aim of growth maximization, money maximization and wealth maximization orients towards the crude exploitation of global human resources and natural resources. This is the fundamental capitalist cause of the earth shaking disparity between rich and poor people and between poor developing and most developed countries of the world. The present global corporate capitalist system operating and functioning through multiple nation-states can be developed into a single global corporate capitalist system with global capitalist state but it will take very heavy cost in the form of death of mankind/life and destruction of whole human civilization on Planet Earth. Now the Planet Earth and majority mankind are not in a position to bear further the burden of global corporate capitalism a ruthless exploiter of nature and mankind.

CHEMICAL BASED GLOBAL AGRICULTURE

Historically the agriculture is 10 thousand years old in human society. After the transformation of feudalism (monarchical kinship) into industrial national formation due to the emergence of industrial scientific-technological mechanism and post-1945 scientific-technological revolution within the corporate capitalist system (national industrial formations) which changed the world of sovereign nations into the world of interdependent nations, has reduced working numbers in agriculture in the developed world. Today the figure is 4% of the occupied population in OECD countries and 2% in the US. In the mid of 1960 there were five nation states in Europe with more than half the occupied population working in agriculture, eleven in the Americas, eighteen in Asia and three were exceptions Libya, Tunisia and South Africa. Today the situation is dramatically very different. There is now no country in Europe or the Americas with more than half the occupied population working in agriculture and same is true in the Islamic world. Even Pakistan has fallen below the 50% mark, while Turkey has moved a peasant population of three quarters to one-third. Even the major fortress of peasant agriculture in south-east Asia has been breached in several places. Indonesia is down from 67 to 44%, the Philippines from 53 to 37%, Thailand from 82 to 46% and Malaysia from 51 to 18%. By 2006 even China with 85% of its peasant population was down to 50% in 1950. In fact, with the exception of most of the sub-Saharan Africa, the only solid bastions of peasant society left (over 60% of the occupied population in the year 2000) are in the former South Asian Empires at Britain and France— India, Bangladesh, Myanmar and Indo-Chinese countries. In the late 1960 the farming population formed half of the population in Taiwan and South Korea. Today it has fallen to 8–10%. We will have ceased within a few decades to be what mankind has been since its emergence as species whose members were chiefly engaged in food gathering, hunting or producing food and goods. Today the mankind is facing a serious multidimensional global agriculture crisis. The global corporate capitalist unsustainable agriculture model based on chemical agriculture technology or inorganic agriculture technology (which was named as green revolution) mainly owned and managed by the corporate capitalists and landlords (including rich and medium farmers) is the fundamental cause of present global agriculture crisis. The chemical based inorganic agriculture technology damaged the natural agriculture factors (the health of agriculture land and its natural pest resistance) due to the use of chemical fertilizers, pesticides, insecticides and herbicides harmed the mankind through the consumption of inorganic agriculture products. The science has proved that the inorganic food products are slow poison, which is terribly harming the human health. It is one of the major causes of deadly diseases. The corporate capitalist policies, continuous rise of agriculture cost of production and the impact of global climate change also seriously affected the global agriculture, especially the poor and medium farmers of poor and developing countries who are the main victims of the global agriculture crisis.

The world is facing more problems, with the climate changing rapidly, in maintaining adequate agriculture production. More challenging than boosting agricultural yields overall is that changing weather patterns means certain areas cannot sustain themselves. The people migrate to cities but the urban infrastructure is unable to support such burgeoning populations. This in turn sows the seeds for social conflicts, which impedes any step towards good governance and actually digging out from a long-downward cycle. About 20 countries of the world are in this condition.

The global food sector has been highly responsible to market forces but farm production probably will continue to be hampered by wrong and misguided agriculture policies that limit investment and distort critical price signals. Keeping food prices down to placate the urban poor and spur saving for industrial investment has distorted agricultural prices in the past. The political elites are more worried about urban instability than rural incomes; their policies are likely to persist, increasing the risk of tight supplies in the future. The world will have to juggle competing and conflicting energy security and food security concerns. Some economists are of the opinion that with global market settling a lower grain volume speculation—invited by expectations of rising fuel costs and more erratic climate change—induced whether pattern could play a greater role in food prices.

The other important factors responsible for the global agriculture crisis are:

1. The corporatization of agriculture sector and continuous focus on corporate industrialization has deprived the farmers from their agriculture land and reduced agriculture area.

2. The building or formation of special economic zones, big commercial complexes, shopping malls, etc. in agriculture land also deprived the farmers from their agriculture land and reduced the agriculture area.

3. The global corporate capitalist system accords first priority to corporate industrial manufacturing sector and service sector and give third position to agriculture sector, a highly important sector of human society (a lifeline of mankind). This wrong approach about agriculture has made the agriculture sector backward in most of the countries of the world.

4. The lopsided and wrong agricultural policies of the national governments of the world and their tilt towards global corporate capitalist system ignored the interest of global agriculture sector and the farmer community.

5. The global agriculture sector is also seriously suffering due to the global crisis of global warming/climate change. The droughts, floods, cyclones, agriculture land erosion, falling of underground water table and uncertainty in climate adversely affecting the global agriculture.

6. The very less investment (public and private) in agriculture sector in comparison to corporate industrial manufacturing and service sectors.

7. The continuous fragmentation of agriculture land in almost all the underdeveloped and developing countries (where the agriculture sector is still unorganized).

8. The lack of level playing field for agriculture competition at global level due to the huge subsidy based agriculture policies of US and developed European countries.

9. The continuously failure of monsoon and deteriorating economic conditions or impoverishment of the farmers are forcing them to migrate or to commit suicides (a great loss of mankind). This is happening in sub-Saharan region of Africa, India and many poor countries of the world. Today one billion (100 crores) people of the world are suffering from hunger as per the latest report of (Nov. 2009) global food conference held at Rome, Italy.

The corporate capitalist governments in the world have called in the very beginning of the emergence of chemical based inorganic agriculture technology as the emergence of the 20th century green revolution. They claimed that the chemical based inorganic agriculture technology is the solution of global hunger and cornerstone of food security for all places and all times. But all these claims have proved wrong and flopped. In fact, after the adoption of chemical based inorganic agriculture technology in agriculture sector throughout the world, the agriculture productivity increased many folds but in spite of this the whole world is confronted with more deadly problems in the sphere of agriculture. The use of chemical fertilizers, insecticides, pesticides, herbicides, etc. destroy all the organic nutrients contained in the soil and kill the whole lot of the crop friendly pests. Thus resulting in the degradation of agriculture soil and finally turning into a barren land. The unsustainable chemical based inorganic agriculture technology is also the cause of emission of green house gases (methane and nitrous oxide) and water pollution. The chemical based inorganic agriculture technology was introduced and produced in the world by the multinational companies of global corporate capitalists. There is a monopoly of corporate capitalist class on chemical based inorganic agriculture technology or agriculture inputs. The pro-corporate capitalist governments all over the world adopted the chemical based inorganic agriculture technology. The corporate capitalist and multinational companies are earning billions of dollars as subsidy on their products (chemical fertilizers, pesticides, insecticides and herbicides) from almost all governments of the world. The negative impact and degeneration of natural resources and human resources has now been accepted by all the renowned agriculture experts of the world including UN Food and Agriculture Organization. Now there is a search of new agriculture model in order to bring the global agriculture out of prevailing multi-sided global agriculture crisis.

After serious negative results of chemical based inorganic agriculture model the scientists and agriculture experts of the global corporate capitalists or multinational companies introduced genetically modified or transgenic agriculture technology in agriculture sector in the world. But the scientific tests show that like chemical based inorganic agriculture technology, it is also harmful and anti nature–anti mankind agriculture technology. The corporatization of agriculture and contract farming by the corporate capitalists is also anti-farmers and anti-people policy. It deprived the poor and medium farmers from the agriculture management and agriculture land ownership. The genetically modified agriculture technology, corporatization of agriculture and contract farming are not the solutions of present global agriculture crisis. The state regulated socialist agriculture model managed through bureaucrats (experienced in socialist-communist countries) which deprived the farmers from agriculture land ownership rights was a one-sided and discarded agriculture model.

The appropriate and rational solution of present global agriculture crisis is to fundamentally change, restructure and remodel the present chemical based inorganic agriculture system and establishes nature-mankind friendly organic-cooperative global agriculture system.

THE POSITIVE AND NEGATIVE FEATURES OF GLOBAL CORPORATE CAPITALISM

The global corporate capitalist system despite its fundamental systemic defects, strategic weaknesses and anti nature–anti mankind basic character; it has some positive features:

1. The capitalism replaced the old monarchical feudal system throughout the world and established capitalist system which is comparatively advance than feudal monarchical system. The capitalists organized the small so many divided feudal territorial states into nation-states, primarily in Europe and then in the rest of the world.

2. The capitalism provides the right of universal suffrage and some democratic rights which were not enjoyed by the people in the feudal age.

3. The capitalism advanced and modernized the scientific-technological mechanism by bringing scientific-technological revolution in the domain of key technologies, i.e. (i) Space technology, (ii) Informational technology, (iii) Biotechnology, (iv) Energy technology, (v) Material technology. These technologies brought a fundamental change in the basic nature of capitalist world which has fundamentally changed the sovereign nations world to interdependent world and boosted the process of globalization and made exclusive nationalism irrelevant. Ultimately the historical process of nationalism will disappear from the global scene and a single global system will take the place of old national systems.

4. The global corporate capitalist system due to advance technology promoted efficiency, knowledge, human interaction, management skills, global trade and commerce, international sports, global communication, global tourism and transportation and productivity many folds comparative to feudal age.

5. The global corporate capitalist system promoted space and extra-terrestrial knowledge or greater knowledge about vast universe.

6. The global corporate capitalist system promoted education including medical education.

But in spite of so many advances and developments done by the global corporate capitalist system in its long historical process, now it has become outdated and irrational, proved anti nature-anti mankind, the creator of the most dangerous fundamental challenge of global warming/climate change and global dehumanization/human cum social inequality. The global corporate capitalist system now has become irrelevant, because it is not conforming to the new global realities, human urges, the demand of systemic change of human society and the need of 21st century. The periodical global economic recession, accumulation and monopolization of capital or wealth in the hands of few corporate capitalist players and the threat of global climate change and global dehumanization are proving the corporate capitalist model an irrational, irrelevant and unworkable social model. The market regulated western corporate capitalist model will not prove sustainable in China. If it does not work for China then it would not work for India and the other 3 billion people in developing countries and former socialist-communist countries, which are also following the path of western corporate capitalist model. The human society needs fundamental change in the market regulated global corporate capitalist system. Today the important global task before mankind is to build a nature–mankind friendly new economic system, which should be powered largely by renewable sources of energy, sustainable technology, and clean raw materials and must have highly diversified transport system that reuse, recycles everything with supersonic speed and managed, controlled and owned by the global people.

5

Socialist-Communist Model

MARXIAN SCHOOL OF THOUGHT

The mankind had a great historical experience of socialist-communist model/system (near about 75 years) and as an alternative school of thought of new social model/system comparative to capitalist model or system. Karl Marx, a great scientific-social thinker of 19th century, challenged the capitalist model/system and free market regulated economic school of thought of Adam Smith. Karl Marx continued and consummated three main schools of thought of 19th century (three main ideological currents) (i) The classical German philosophy (ii) The classical English political economy (iii) The French socialism combined with French revolution doctrines. The Karl Marx (with Engels) after the deep study of capitalism, historical process of human society, the Hegelian dialectical idealism, the Hegelian socialism, the classical capitalist political economy and experience of French revolution, etc. discarded capitalist theory, capitalist model, idealistic philosophy and religious myth or faith. He propounded a new scientific school of thought based on materialistic approach, i.e. dialectical materialism, historical materialism and alternative social model of scientific socialism-communism. The Marxian school of thought and scientific socialist-communist model reflected in these theories:

1. The dialectical materialism.
2. Historical materialism.
3. Theory of class struggle.
4. Scientific socialism-communism.

5. Theory of capital and surplus value.

6. Theory of proletarian revolution, etc.

All these theories are expressed in Marx's famous writings of "communist-manifesto", capitalist classical political economy, Das Capital (in three volumes), dialectical materialism, the evolution of family, private property, classes and state, role of force in history, the proletarian revolution, etc.

Karl Marx analysed the exploiting nature of capitalist system, fundamental contradiction of capitalist system, concept of capital, source of capital and inevitable solution of capitalist contradiction through proletarian violent revolution and the establishment of scientific socialism-communism (in his famous writings Das Capital, political economy, communist manifesto, etc). He holds that in capitalist system there is a mode of production (base of capitalist system) and superstructure (including capitalist state powers). The mode of production is constituted by the productive forces and productive relations. Further he analysed that the character of productive forces are social but the nature of productive relations (ownership and control) are private and capitalist. The fundamental dialectical contradiction of capitalist system lies in the capitalist mode of production (capitalist base) between social productive forces and private/capitalist productive relations, ultimately reflects in the capitalist superstructure in the form of class struggle between proletariat (the modern working class) and bourgeois (the capitalist class). The contradiction in the capitalist mode of production and class struggle between proletariat class and capitalist class in superstructure creates inevitable conditions of proletarian violent revolution and the solution of contradiction of social productive forces and private/capitalist relations (ownership and control) into social relations (social ownership and control) in the form of scientific socialism-communism which is historically predetermined in capitalist system's contradiction. When private ownership will be abolished, the classes will be abolished and then there will be no need of class state power. Ultimately after the abolishment of private property, the classes and state power will wither. There will be classless and stateless society, the highest stage of human society will be called communism, in which the principle of according to ability and according to need will be observed.

Karl Marx has given supreme importance to the concept of capital while differing with the capitalist concept on source of capital but he too has defined capital in terms of money. Marx holds that the capital comes from the workers surplus value. The capitalists exploit the workers by paying them less than the due money wages. The capitalism considers that the capital arises during production process, capital investment, technology, labour and raw material with capital owners playing the main role. In fact, capital theory of Marx focus only on labour as the source of capital while the Corporate Capitalist theory emphasizes all the main abovesaid factors are the source of capital but gives central position to the capital owner. The Marxian Communist theory analysed that the capitalist system is the root cause of workers exploitation, social inequality, poverty, accumulation and monopolization of capital around the capitalists. He propounded a new social model of scientific socialism-communism, a model based on social equality and free from capitalist exploitation.

POSITIVE AND NEGATIVE HISTORICAL EXPERIENCE OF SOCIALIST-COMMUNIST MODEL/SYSTEM

The communist model of Karl Marx inspired the workers, toiling masses and intellectuals of the world. He gave the slogan of proletarian revolution and workers of the world unite. The large scale communist parties and revolutionary movements emerged throughout the world on the basis of Marxism and Communist model, which culminated into armed revolution in so many under developed countries instead of advanced capitalist countries as predicted by Marx. The Marxism developed into Leninism by Lenin, which reflects in his writings, dialectical materialism (unity of opposite), the left wing communism infantile disorder, the state and revolution, one step forward two steps backward, imperialism the highest stage of capitalism (the moribund, decaying and parasitic capitalism), the state and revolution, concrete analysis of concrete conditions is the soul of Marxism, Marxism and revisionism, developed the concept of vanguard proletarian revolutionary party and proletarian dictatorship, the concept of War Communism, the New Economic Policy (NEP 1921), etc. Lenin was fundamentally a planner and man of action. In theory he was a fundamentalist but in practice he modified Marxism. The first proletarian armed revolution occurred in October 1917 in Russia with Marxist Bolshevik Party under the leadership of Vladimir Lenin. In 1917 Lenin whose hopes had not gone much beyond bourgeoisie democratic Russia in 1905, also concluded from the start that the liberal horse was not runner in the Russian revolutionary race. However, in 1917 it was as clear to him as all other Russian and non-Russian Marxists that the conditions for a socialist revolution were not present in Russia. Lenin's extraordinary achievement was to transform uncontrollable anarchic popular surge into Bolshevik power. The basic demand of the 80% Russian, who were living by agriculture, was for land. The basic demand of poor Russian was for bread, better wages for workers and shorter working hours. The masses wanted peace and end of war. The slogan of bread, peace and land won rapidly growing support for those who propagated them notably Lenin's Bolsheviks who grew from a small group of a few thousands in March 1917 to a quarter of a million members by the early summer of that year. Lenin essentially as an organiser of revolution, the only real asset he and Bolsheviks had the ability to recognize what the masses wanted, to lead by knowing how to follow. For instance when recognized that contrary to the socialist programme, the peasants wanted division of agriculture land into the family farms he did not hesitate for a moment to commit Bolsheviks to this form of economic individualism. The peasantry gave overwhelming support to the heirs of the Narodniks and the socialist revolutionaries. Though they developed a more radical left wing which drew closer to the Bolsheviks and joined them in the government after 1917 October revolution. The Bolsheviks found majority in major Russian cities, especially in capital, Petrograd and in Moscow and gained ground in army. The provisional government's existence becomes increasingly shadowy. In fact, when the moment came power had not so much to be seized as to be picked up. The provisional government with no one left to depend it, merely dissolved into thin air. From the moment the fall of the provisional government become certain the October revolution drenched in polemics. Most of them were misleading. The real issue was not whether it was a coup by Lenin but who or what should or could follow the fall of the provisional govt. In the early September 1917, Lenin tried to convince the hesitant elements in his party that not only the power might easily escape from them if not seized by planned action during the possible short time when it was within their grasp. But to answer the question, 'can the Bolsheviks retain state power if they seized it'. No party other than Lenin's Bolsheviks was prepared to envisage this responsibility on its own. Lenin was fully determined for the seizer of political power. In spite of wavering and vacillation of some Bolsheviks Lenin suggest that the given favourable situation in Petrograd, Moscow and the northern armies, the short-term case for seizing political power now, rather than waiting further on events, was indeed difficult to answer. Lenin rarely hesitated to look the darkest facts in the face. He said, if the Bolsheviks failed to seize the moment 'a wave of real anarchy may become stronger than we are'. In the last analysis Lenin's argument convinced his party. If a revolutionary party did not seize political power when the moment favourable and masses called for it, how did it differ from a non revolutionary one? At last the Bolshevik party under the leadership of Lenin pushed the final assault. The Bolsheviks party of Russia seized state power from Czar and built first socialist state and nationalised all the means of production of Russia. The Bolsheviks not only maintained their power longer than Paris commune of 1871 but through years of unbroken crisis and catastrophe. The Russian October 1917 revolution inspired not only revolutionaries but more important revolutions. The Russian revolution survived because of three major reasons. First, it possessed a uniquely powerful a state building instrument in the 6,00,000 strong, centralized and disciplined communist party of Soviet Union. Second, it was quite evident that the government under Bolsheviks was able and competent to hold Russia together as a state and therefore enjoyed considerable support from political hostile patriotic Russian. The Bolshevik revolution and CPSU preserved most of the multinational territorial unity of old Tsarist state at least for another seventy-four years. The third reason was that the revolution under the guidance of Lenin allowed the peasantry to take the land. When it came to the point, the bulk of the Russian peasants thought their chances of keeping the land better under the Bolsheviks. This gave the Bolsheviks a decisive advantage in Civil War of 1918–20; it turned the Russian peasants too optimistic. The world revolution which justified Lenin's decision to commit Russia to socialism did not take place and with it the socialist Soviet Russia was committed to a generation of impoverished and backward isolation.

Initially Lenin introduced the policy of war communism (total abolishment of private ownership) but very soon he realized the failure of economy under war communism and he withdrew the policy of war communism and introduced the revised policy of new economic policy of Russia (NEP, 1921). In fact, Lenin reintroduced the market and retreated from war communism to state capitalism. The radical followers of Trotsky wanted break with new economic policy. The moderates led by Bukharin were keenly aware of the political and economic constraints under which the Bolshevik government had to operate in a country overwhelmingly dominated by agriculture than before the revolution. The Bukharin favoured a gradual transformation. But Lenin's views could not be adequately expressed after paralysis attack to him in 1922 and he survived until early 1924. While he could express himself, he looked to favoured gradualism. The important fact of Soviet Union was that its new ruling Bolshevik party had never expected to survive in isolation, but alone to become the nucleus of a self-contained collective economy (socialism in one country). None of the conditions which Marx had hitherto considered essential for the establishment of socialist economy, were present in this enormous country, which was virtually a synonym for economic and social backwardness in Europe. The founder of Marxism assumed that the function of Russian revolution could only be to spark off the revolutionary explosion in the more advanced industrial countries where the preconditions for the construction of socialism were present. This exactly happened in 1917 and appeared to justify Lenin's highly controversial decision among Marxists to set the course of the Russian Bolsheviks for Soviet power and socialism. In Lenin's view the Moscow would only be the temporary head quarter of socialism until it could move to its permanent capital in Berlin. The official language of the communist international was not Russian but German. But at that time the Soviet Union was the first country in which the proletarian revolution had triumphed and first socialist-communist system was established. After the October 1917 revolution the Bolshevik party turned from an underground organization of few thousand illegal Marxist comrades into a mass party of millions of professional mobilizers, administrators, executives and controllers, who swamped the old Bolsheviks and other pre-revolution socialists who joined them, such as Trotsky. They did not share the old political culture of the left but all they knew was that the party was right and the decisions made by the superior authority must be carried out if the revolution and socialism to be saved. The ruling Bolshevik party formulated the policy of Soviet Union to transform its backward economy into an advanced economy. The building of Soviet Union based socialist-communist system became the primary programme for transforming backward countries into advanced ones. After the death of Lenin (1924), Stalin took the reign of Soviet Union, in spite of Lenin's refusal to back Stalin as his successor because of his excessive brutality. The Stalin with dictatorial and tyrannical methods started the building of socialism in Russia under his theory of socialism in one country. He brutally suppressed and crushed every dissension and resistance from the leaders, party activists, intellectuals, workers, peasants and any shade of opinion from the people during the period of building the socialism and land collectivization. The monopolization of the means of production by socialist state deprived the people from their ownership rights and led it towards the single party state regulated state-economy and one party political dictatorship in Soviet Union. It was neither social productive relations (social ownership and control) nor proletarian dictatorship as propounded by Karl Marx and Lenin. The slow down of the Soviet economy was palpable. The rate of growth of almost everything (gross domestic product, industrial output, agriculture output, capital investment, labour productivity, per capita income, etc.) fell steadily from first five-year plan to next after 1970. The economy was advancing at the very slow pace. In fact, by the 1970 it was clear that not only economic growth was lagging but even the basic social indicators were ceasing to improve. This undermines the confidence in socialist model/system. In fact, it was crude and cruel totalitarian form of social model/system. The Brezhnev period in Soviet Union was to be called the era of stagnation, because the Brezhnev regime had stopped trying to do anything serious about a visibly declining economy. It adopted the path to buy wheat from world market then trying to cure the growing inability of Soviet agriculture system. The historical experience and the nature-mankind friendly/centric model guide us that the proletarian dictatorship or one party dictatorship or totalitarian state regulated socialist model is an irrational social model which has proved contrary to the interests of nature and mankind.

The Soviet Union (after the second world war) under its influence and support brought—Czechoslovakia, Romania, Hungry, Poland, Bulgaria, and East Germany—in the fold of socialist model. The socialist system also established in Yugoslavia under the leadership of Marshal Tito. But Marshal Tito took independent line from the Soviet Union. The Albania also became a socialist state after World War II. The socialist-communists consolidated their power in eastern Europe after the World War II.

The communist party of China under the leadership of Mao Zedong brought armed revolution against Chankey Sheik regime and seized political power of China in the year 1949. The communist party of China established its dictatorship, nationalised all the means of production like Soviet Union, abolished the private property and established state power regulated socialist model with the tune of socialist model of Soviet Union. But in fact, unlike Russian socialist model, the Chinese socialist model had virtually no direct relations with Marxism. It was a post-October revolutionary movement which came to Marx via Lenin or it was Stalin's Marxism-Leninism. Mao's own knowledge of Marxist theory has been almost entirely derived from the Stalinist history of CPSU. Mao's intellectual development was fully home grown and he seized on those aspects of Marx and Lenin which fitted into his vision. His view of an ideal society was collectivist mysticism, the opposite of classical Marxism, which at least in theory and as the ultimate object, envisaged the complete liberation and self-fulfilment of the individual. The communist party of China under the leadership of chairman Mao divided the whole China into 24,000 people communes, adopted close door policy and delinked China from the rest of the world except socialist countries. The 84% of Chinese peasant households had allowed themselves to be collectivized within a single year 1956, without any of the consequences of Soviet collectivization. The industrialization on the basis of heavy industry reflected Soviet model. Unlike the USSR the China experienced no mass urbanization under Mao. The per capita economic growth of China in the period of Mao was less than that of Japan, Hong Kong, Singapore, South Korea and Taiwan. In the year 1965, Mao with military backing launched an anarchic movement of young red guards against the party leadership which had quietly sidelined him. This was the great cultural revolution which devastated China for some time, until Mao called in the people liberation army to restore order and to restore party control. The Mao thought without Mao had very little real support. After the death of Mao and arrest of the gang of four headed by the widow of Mao Jiang Quing, the new course under the pragmatist leader Deng Xiaoping began immediately. Deng Xiaoping took the reign of China in the year 1978 after the death of Mao. He remodelled the socialist model of China established by Mao (except communist party of China regulated old authoritarian political setup) and started the corporate capitalist modernization of China. He opened and integrated the Chinese socialist economy with global corporate capitalist market and global corporate capitalist system. China fully invited the global corporate capitalist multinational companies (on the basis of 100% equity or profit, to which they can bring back in their nations or elsewhere) particularly the US and western world's companies (to whom Mao once said, they are exploiters and paper tigers). Now China has become a big global corporate capitalist player, a competitor of the US (a global corporate capitalist superpower). Deng Xiaoping said, "the key to achieving modernization is the development of science and technology. The empty talk will get our modernization programme nowhere; we must have knowledge and trained personnel". Now it appears that China is fully twenty years behind the developed countries in science, technology and education. As early as the Meiji restoration the Japanese began to expend a great deal of effort on science, technology and education. The Meiji restoration was a kind of modernization drive undertaken by the emerging Japanese bourgeoisie. As proletarians we should and can do better. The China a socialist country in 1970s was particularly worried by its relative economic backwardness, because its neighbour Japan was the most spectacularly successful capitalist country. The technological inferiority of China was not due to technical or educational incapacity, but to the very sense of self-sufficiency and self-confidence of traditional Chinese civilization. This made it reluctant to do what the Japanese did after the Meiji restoration of 1868: plunge into modernization by adopting European models wholesale.

Cuba, North Korea, Cambodia, Vietnam, Laos, etc. also established state regulated socialist systems. The socialist-communist states such as Soviet Union and China succeeded to become industrial, technological and military powers to compete the capitalist powers especially US and NATO countries in arms race, space race and military conflicts during the period of cold war between US and USSR. The one-third of the world population lived in the socialist-communist states including former Soviet Union and China by the early 1980s. The people deprived from the individual ownership rights, democratic rights, freedom of expression and freedom of dissension in the socialist-communist ruled countries. Nobody was allowed to point out the weaknesses of the socialist-communist system and their policies. The socialist-communist system adopted dogmatist approach. The diehard communist claimed that their socialist-sommunist system is an absolute truth. There is no scope of any change in the socialist-communist model/system. The capitalist model continuously brings step by step some reforms within the framework of capitalism. But the socialist-communist model restrained by the mythical understanding of its being the repository of the "absolute truth", failed to perceive the new global realities. Thus persisted with their traditional mode of model, thoughts, actions and charactering every demand for even a slight change in their model considered as outright revisionism or against unchangeable socialist-communist model.

In fact, every social model is subject to change with the tune of continuous changes in natural and social processes. The change and development through the law of interaction between nature and mankind and within mankind is a regular feature of social reality, which was overlooked by the advocates and followers of socialist-communist model.

Karl Marx stands for the socialization of the means of production on the one hand and workers centric approach on the other hand. This was a contradictory and confused stand. Moreover Marx did not comprehensively propound that how to build socialism-communism except he only mentioned that to make productive relation social according to the social productive forces. Those who implemented Marxism after revolutions, e.g. Lenin in Russia, Mao in China, Kim il Sung in North Korea, Fidel Castro in Cuba, Anwar Hoxa in Albania, etc. have nationalised all the means of production and made state ownership, control and regulation instead of people's democratic social ownership (ownership of the society) and in politics established one party dictatorship instead of people regulated democracy or people governance.

The Marxian theory takes a confusing stand with regard nature of mankind. On the one hand it characterizes that mankind is social by nature and on the other hand it says that the industrial proletariat (modern industrial worker) is the liberator of the mankind and will play leading role in the revolutionary change. But in practice the Marxist Socialist-Communist Model transferred the entire power to the communist party and party transferred the powers to politburo and party general secretary. This undemocratic and authoritarian model established the totalitarian dictatorship of party and party leaders. The socialist-communist model holds that the communist party has capacity to bring social change/transformation in human society. This approach ultimately monopolized and centralized the whole authority and power in one person, i.e. the general secretary of the party who remain the dictator for whole of his life till death. In Soviet Union less than one party rule, the communist party of Soviet Union (CPSU) particularly its politburo monopolized whole power and state regulated socialist system, which ultimately changed into the totalitarian and the most centralized system of the world (with the absence of healthy political and economic competition, freedom of expression, democratic rights, human rights and individual property rights of the people). The other socialist-communist countries of the world (China, North Korea, Cuba, Albania, Romania, Hungry, Bulgaria, East Germany, Vietnam and Cambodia) also adopted the same totalitarian approach. After the long process and practice (75 years) of state regulated socialist-communist model/ systems the degeneration started in these systems due to their one-sided and unreal concepts, unscientific and dogmatic approach, one-sided law of contradiction or class struggle, one-sided workers centric approach, one-sided inhuman approach of violent revolution, proletarian dictatorship and absolute state ownership of socialist economy. The socialist-communist countries also established totalitarian-dictatorial state and state regulated bureaucratic managed so-called socialist economies instead of people ownership, democratic people management and democratic regulation of economy by the whole people of human society. The state regulated socialist-communist model (through top politicians and bureaucrats) produced new dominant classes in all these socialist countries.

The serious systemic defects, wrong understanding of capital, capital or money centric approach of socialist-communist systems, similar to corporate capitalist model, orientation of socialist-communist countries towards growth maximization and money/capital maximization proved that there is no fundamental difference between market regulated corporate capitalist model/system and state regulated socialist-communist model/system. The only difference between states regulated socialist-communist model/system and market regulated corporate capitalist model/system is that the corporate capitalist model/system is mainly regulating through market mechanism and the socialist model/system is mainly regulated through state power. The socialist-communist state power was controlled and regulated by top communist leaders and top bureaucrats. The socialist-communist politicians and bureaucrats became a new dominant ruling class. Whosoever tried to dissent or oppose the socialist-communist system was ruthlessly suppressed by the socialist-communist political establishment. The socialist systems confronted not only their own increasingly insoluble systemic problems but also those of changes and problematic corporate capitalist world economy into which they were integrated.

Mikhail Gorbachev took over the reign of socialist-communist Soviet Union in the year 1985. He started democratic reforms in Soviet Union (USSR) under the agenda of Perestroika (restructuring of Soviet Union's old socialist system), Glasnost (openness of Soviet society from a close society) and Democratization of Soviet Union. The Mikhail Gorbachev through dialogue with US superpower (during Ronald Regan period) ended superpower cold war. He dissolved the Warsaw Pact (a socialist military pact), started restructuring of the socialist system, dissolved the socialist block, encouraged the democratic changes in the socialist countries of eastern Europe and facilitated the fall of Berlin wall and east-west Germany's merger with the co-operation of German Chancellor Helmut Kohl (a great historical event). The east Germany becomes the part of corporate capitalist model of west Germany. Almost all the countries of the socialist block adopted the system of multi-party democracy and market regulated corporate capitalist model.

The Perestroika, Glasnost and Democratic reforms initiated by the Mikhail Gorbachev in Soviet Union, were heartily welcomed by the Soviet people and his contribution to end the super-powers cold war with US was also appreciated by the world people. But Gorbachev's hesitant style and his hob-nob with hardliners of Communist Party of Soviet Union and sidelined by him the democratic leaders, created confused state of balance of power in Soviet Union. The hardliners had relatively strong position in the Communist Party of Soviet Union (CPSU) and government hierarchy. The hardliners did not allow the Gorbachev's reforms with human face to move towards their logical end. Gorbachev wanted to change the Communist Party of the Soviet Union (CPSU) into social democratic party and Soviet Union's totalitarian model into federal system in a special congress of Soviet Union but it was sabotaged before special congress (which could not be held) through coup by the hardliners. The Gorbachev's proposal to form social democratic party and building of federal Soviet Union was not the right answer and not the alternative model for the old socialist system. The market regulated global corporate capitalist system is a new and highest global stage of capitalism. In fact, Gorbachev lacks a clear-cut and comprehensive alternative social model (distinct from market regulated global corporate capitalist model). His social model does not corresponds with the new global realities and nature-mankind friendly/centric spirit. The internal and external systemic structural changes initiated by the Gorbachev changed the global balance of power due to the withdrawal of 'Soviet Union's armed forces from the Afghanistan, Vietnam's armed forces from Cambodia, merger of east and west Germany, end of superpower cold war, dissolution of socialist block and Warsaw pact, openness of Soviet Union, restoration of democratic rights of the people of Soviet Union and other socialist countries of the socialist block.

But unfortunately due to the systemic defects of the Soviet socialist system, coup of diehard communist leaders of CPSU (along with some officials), political mistakes committed by Gorbachev, sectarian or narrow Russian sub-nationalist line or opportunist line of Yeltsin, cold and hot war of US and NATO countries against USSR, wrong advice of Yeltsin to Gorbachev to dissolve the Communist Party of Soviet Union (CPSU) and accordingly the dissolution of the Communist Party of Soviet Union by Gorbachev without forming an alternative organization (it was the biggest blunder committed by both Yeltsin and Gorbachev) led to the disintegration of USSR and the collapse of 74 years old socialist system of Soviet Union. The Soviet System failed because there was no two-way traffic between those who took decisions in the interests of the people and those on whom these decisions were imposed. It was an unprecedented historical development and social change (after the disintegration of Yugoslavia) seems in the post-second world war period. The Soviet Union a 45-year-old super power with 27,000 nuclear warheads, having biggest army in the world and Gorbachev's human face model could not save the Soviet Union from disintegration and collapse of totalitarian socialist system. The contrast between Soviet Union and Chinese restructuring process was that the Chinese were careful to keep their central command system intact, while in Soviet Union the Gorbachev weakened it. The US along with NATO Block declared the defeat of socialist model and claimed victory of market regulated global corporate capitalist model over the communist model. It is the point worth for consideration that Gorbachev in the process of dialogue with US super power ended cold war and dissolves the Warsaw military pact. But why the US and the western block continue the NATO and the super-power policy of US, which is counter-productive, anti nature-anti mankind and contrary to the global historical need of harmony and solidarity between nature and mankind and within mankind.

The historical experience of Marxism and socialist-communist model is having both positive and negative aspects. The positive side provides many useful and beneficial achievements. While its negative side contain very serious failures which led to its eclipse from the competition with its rival Adam Smith's capitalist theory and global corporate capitalist model/system in the realm of human development. The socialist-communist model/system discarded throughout the world. The history refuted the socialist-communist model as a one-sided and unreal model. But still there are Marxists in the state of decline and confusion. In spite of discarding communist model by time and history, its deep theoretical and vast positive and negative experience can be useful for further study and research, to formulate nature-mankind friendly new global confederal model, because both the global corporate capitalist and the communist models proved one-sided, outdated, irrelevant and anti-nature and anti-mankind. The unworkability of state regulated socialist-communist model proved from the collapse of Soviet Union, corporate capitalist changes occurred in China and the whole socialist block countries.

The history tells us that it was a great and positive attempt by Karl Marx, Lenin, Mao, Fidal Castro, Kim il Sung, Anwar Hoxa, Marshal Tito and other revolutionary leaders, communist activists, workers, peasants of the world and communist parties of the world, to build a socialist-communist systems in order to bring social equality and to end capitalist exploitation. But unfortunately due to wrong understanding of nature and mankind, one-sided concepts and approaches, workers centric approach, dogmatist attitude about reality, authoritarian and totalitarian political setup, absence of democratic and individual rights of the people, systemic serious weaknesses and adherent to dogmatist approach, i.e. not to learn lessons from the past mistakes and negative experiences, ultimately pushed the socialist-communist systems into utter failure. Mr Lewin in Kerblay 1983 said, 'it is not true any more that a single official creed is the only operative guide to action. More than one ideology, a mixture of modes of thinking and frames of references, co-exist and not only in society at large but also inside the party and inside the leadership. 'A rigid and codified 'Marxism-Leninism' could not, except in official rhetoric, respond to the regime's real needs.

In the globalized corporate capitalist interdependent world mainly Cuba, North Korea and Vietnam still carrying the dogmatist and one-sided Marxist line and state regulated socialist system which has already become irrelevant and unworkable today. But now in the socialist Cuba the large scale discussion among the masses has been initiated by the ruling party and socialist govt about the reduction of govt role in economy, dismantling of state regulated socialist system and promotion of privatization. The Cuban ruling party and govt is seriously seeking the opinion of Cuban people in this matter. Keeping in view the failure of both the socialist model and corporate capitalist model, the people of Cuba should express their democratic opinion in favour of nature-mankind friendly global confederal system instead of market regulated corporate capitalist model. The experience of former communist countries is that on the one side the people overthrown the communist dictatorship in most of the former communist countries but due to lack of concrete new alternative model they established outdated and anti nature-anti mankind market regulated and multi-party regulated global corporate capitalist model. The communist countries that become under the influence of global corporate capitalist system and adopted market regulated corporate capitalist model sooner or later they will realize their great mistake and will be bound to correct their historical blunder they have committed in history. The new global realities and the historical need of global fundamental social change in global corporate capitalist system will defiantly convince and motivate the global people or global mankind to adopt nature-mankind friendly global confederal model.

6

Lessons of Corporate Capitalist and Communist Model/System

The historical experience of market regulated corporate capitalist system from the year 1750 (when industrialization started) to onwards and more than 74 years experience of state regulated socialist-communist system testified the fact that both the systems are one-sided, anti-nature and anti-mankind. The human face theory of corporate capitalism is only a wishful thinking. It has never established or appeared in any country or region of the present world. The socialist-communist system of social equality and justice collapsed due to dogmatist and one-sided approach. Today the mankind is living under an unfair and unjust global corporate capitalist social system. In the global corporate capitalist system the powerful dominates the weak and haves ride over the have not with money, might and privilege both on global and on national levels. The global corporate capitalist system and its politicians degenerated the party regulated traditional democratic systems by adopting the money oriented political culture. The capitalist system for attaining its objective of money and power has pursued ruthless colonialism for nearly 200 years the world over, waged two aggressive and inhuman world wars, 40 years deadly cold war the two super powers between US and USSR, so many national wars and civil wars within the nations. The global corporate capitalist system and its governments are the creator of global terrorism. The global terrorism is the outcome of global corporate capitalist system and its policies.

In Tunisia on 19th December 2010 the self immolation of Mohammad Bouazizi, a 26-year-old unemployed, university graduate become the catalyst that sent shock waves throughout the Arab world and Africa. Unlike the stubbornness of President Hosni Mubarak of Egypt, Ben Ali realized the anger of the masses and handed over power to vice president. The dictators of Arab world continued the cruel dynast rules and groomed their heir apparent from among their family. All the dictatorial regimes of these regions resisted to any social, political and economic change. But the effect of democratic and peaceful mass movement for democracy and freedom, spread across the length and breadth of Africa and middle East like a wild bush fire, where the so-called rigged election or ballot boxes failed to bring the desired political change. But the peaceful and democratic great mass movement and mass uprising has proved very powerful and effective means to change these corrupt and dictatorial regimes, than Bullet or so-called elections. The recent mass uprising against dictatorial regimes and movements for freedom and democracy in Tunisia, Egypt, Yemen, Bahrain, Algeria, Libya, Syria, Iran, etc. indicating the historical need of global social change. In these countries the US under its narrow national interests since long overlooked the misgovernance, oppression and suppressions of the people by the dictators and sufferings of the masses due to grave poverty, inequality, unemployment, rising prices, corruption and malnutrition, which has created great mass resentment and mass movements for political change. In Tunisia, Egypt, Yemen, Algeria, Jordan and the rest of the Arab world the masses are so suppressed, crushed and frustrated from the corrupt and tyrant leaders and their poor performance to deliver the basic necessities of life such as employment, social service, health, education and a better and hopeful future. The people under these dictatorial regimes and systems, supported by US and west, suffered a lot from unemployment, rising prices of food products, corruption and tyrant suppression by these regimes. They ruthlessly strangulated the democracy and people freedom. The champion and advocate of democracy the US always supported these tyrant regimes and remained silent about this crime against humanity. The US supported dictatorial countries, particularly the Arab world countries facing similar great mass resentment, mass uprising and movements for freedom and democracy as previously faced by the communist regimes.

The positive aspects of the present democratic political changes in Arab world is changes of tyrannical and most corrupt regimes through peaceful and democratic means without big bloodshed like in Russia (1917), China 1949 and other armed revolutions and created the condition of people democracy and freedom. But the negative aspect of these movements is the lack of realistic and comprehensive alternative model in the tune of present global reality. The present world needs a nature-mankind friendly global confederal system which address the fundamental challenge of global warming/climate change and global dehumanization/global inequality and balance the imbalance state of nature and mankind which is the very essential and necessary factor for the sustainability and survival of human society, human civilization and life. I am optimistic that the global people in general and the people of Arab world in particular very soon will sincerely realize this historically necessary basic global aspect of future global political systemic social change.

Now the great mass movement forced the whole world including US and West to support the movement for political change and forced the dictators to vacate the political seat for masses. The process of global mass movement against individual dictatorship, one party dictatorship, two party dictatorship or multiparty dictatorship will continue. In fact the dictatorial and party regulated so called democracy has become outdated and proved anti nature-anti mankind. The history and new era demand the establishment of global people regulated new global democracy on the basis of nature-mankind friendly global confederal model. In which there will be a global confederal state and all the nation-states should be made confederating units of global confederal state.

Now the time has come in which all the tyrannical and dictatorial regimes will be overthrown in the whole world. In the passage of time the party regulated monopoly regimes or so called corporate capitalist democracy (one party or two party or multi party) will also face the similar situation of global mass uprising and global mass revolution for the replacement of traditional and outdated party regulated democracy and the establishment of people regulated global confederal democracy. Because the global fundamental social change (sovereign nations world into interdependent nations world) has made the party regulated national democracy outdated and created the historical need of global people regulated global democracy on the basis of nature-mankind friendly global confederal model, in which instead of some parties, politicians and bureaucrats to govern and regulate the human society, the whole mankind will become the master, regulator and controller of the every process of global human society. The rationality and historical cum futurist wisdom demands that the mankind should consciously dismantle and dilute the old model/system of party regulated centralized democratic national set ups and establish nature-mankind friendly global confederal democratic single system, with the active participation and involvement of global people. The building of nature-mankind friendly one world system will take place step by step.

The state centric model and market centric model both are one-sided, anti-nature and anti-mankind. The state-centric model in which the state plays in decision key role (in case of China and Russia the democracy is limited) raises the inevitability of the traditional western recipe of liberal economies and democracy for development. There is a possibility that some more developing countries may attract towards the China's state centric model rather than the traditional market centric western model and party regulated democratic political system to increase the chances of rapid development and political stability. But the historical experience tells us that both the state-centric and market-centric are one-sided, anti-nature and anti-mankind. The party regulated democratic system has also become outdated and irrelevant. Neither the state regulated institution/mechanism and nor the market institution/mechanism are the supreme engine and fundamental factors of development. The historical experience of both the models under global corporate capitalist system guide us that neither state nor market is the fundamental factors of development but the nature and mankind are the equally fundamental factors of development and human society. The historical experience of 250 years of corporate capitalism and new realities of human society demands that the present interdependent global corporate capitalist system should be remodelled on the basis of nature-mankind friendly global confederal model. In which the nature-mankind friendly state should be built and all the nation-states should be made the integrated confederal part of global state and global confederal system. The national and international markets should be restructured in accordance with nature-mankind centric/ friendly model. The attraction of some developing countries towards Chinese state centric model is not durable. It will prove wrong like the adoption of market centric model by some former communist countries.

The long experience of market regulated global corporate capitalist model/system and state regulated socialist-communist model/system proved that both the traditional models have become irrelevant and lost their validity and workability. There are defective philosophical approaches and one-sided concepts in both the models. Qystein, the former vice president of Exxon for Norway and the North Sea has observed that, the socialism collapsed because it did not allow the market to tell the economic truth and the global corporate capitalism may collapse because it does not allow the market to tell the ecological truth. The building of socialism in one country and to made one national communist party or one communist country as the global centre was contrary to the international spirit of Marxism. The failure of both the models to maintain harmony and balance between nature and mankind and within the mankind has created the global threat to the survival of mankind and Planet Earth in the form of global warming/climate change and dehumanization/human cum social inequality (global mass poverty, hunger, unemployment, etc.). Both the models have same wavelength and approaches, that capital denotes money and vast sums of money and wealth. That is why both the communist model and global corporate capitalist Model define money or capital generation, growth maximization and capital expansion as the defining feature of the social development, prosperity and progress and gives secondary place to nature and mankind. The state regulated socialist-communist model and the market regulated global corporate capitalist model does not corresponds to the emerging new global reality of interdependence of nations or interdependent world. The global need of harmony between nature and mankind, harmony and solidarity within mankind and growing process of globalization needs nature-mankind friendly new global confederal system. A system governed, owned, regulated and managed by the global people or global mankind.

7 Nature-Mankind Friendly Global Confederal Model

NATURE-MANKIND FRIENDLY PHILOSOPHY

The word Philosophy is derived from the Greek word "Philospia" means 'love of wisdom'. There are so many philosophies in the world but two main broad categories of philosophies exist in the world are as under:

1. The philosophy of faith,
2. The nature-mankind friendly scientific philosophy.

The philosophy of faith believes that there is a supernatural force which is the creator, regulator, operator and destroyers of the universe. The nature-mankind friendly global confederal model upholds the nature-mankind friendly scientific philosophy. The nature-mankind friendly scientific philosophy means "a branch of fundamental human knowledge which explains the existence (in any form or structure), motion and fundamental law of change and development of any phenomenon of the universe (natural, human, bio and social). The nature-mankind friendly scientific philosophy is based on the scientific facts, logics, understanding of the natural sciences and rational social sciences about nature and mankind during the process of observation, experimental investigation and theoretical explanation of the phenomenon or the processes of nature and mankind. The nature-mankind friendly scientific philosophy stands for scientific rational approach, which studies and interprets the natural phenomenon in the light of scientific facts and the process of human society on the basis of authentic data and information of natural sciences and rational social sciences. The nature-mankind friendly scientific

155

philosophy upholds that the two-sided interaction between nature and mankind and within mankind (through alternate primary role of unity and struggle) is the fundamental law of change and development of human society. The scientific way to discover the reality is a real and rational way and correct method to gain the realistic and greater understanding of nature and mankind and human society. The nature-mankind friendly scientific philosophy (on the basis of study and research of natural sciences and rational social sciences) says that the deep scientific study, investigation and research did not find any supernatural force in the universe till now, who is the creator, operator, regulator and destroyer of the whole universe. In fact, the nature works on its own natural laws of existence, motion and change. The great international scientist and physicist Stephen Hawking said in his new book 'The Grand Design' (2010), that supernatural force did not create the universe. The Big Bang was a natural event. There is a law of gravity, the universe can and will create itself from nothing. The Stephen Hawking in his latest book, ''The Grand Design' contradicted his own previous stand expressed in his earlier book 'A Brief History of Time '(1988), if we discover a complete theory it would be the ultimate triumph of human reason—for then we should know the mind of God but Stephen Hawking recently changed his mind. He has given circumstantial evidence that no supernatural force is in control of the universe or multiverse and he tackled in 'The Grand Design 'the question surrounding, the origin of the universe and the nature of the laws of physics, examined how and why the universe was created. The common people have the tendency to think that such questions eternally belong to God and religions. They are not accessible to science. However, according to theoretical physics this opinion is incorrect. The science is doing very deep investigation and research about the origin, creation, very beginning of the existence, motion, change and development of the universe. The Albert Einstein searched for a unified theory which could describe all the forces of nature in a single framework. But the time was not right for such a discovery in Einstein's days. It was also not the right time when Stephen Hawking wrote in 1988 'A Brief History of Time' in which he took us on a journey through classical physics, Einstein's theory of relativity, quantum physics and string theory in order to explain the universe. But now the scientist Stephen Hawking and renowned science writer Leonard Mlodinow in their latest book, 'The Grand Design' a ground breaking new work, based and drawn on forty years research of Stephen Hawking and recent series of extraordinary astronomical observations, focused on reason for the Big Bang and theoretical break-throughs to reveal an original theory.

Humans are curious species. There is multitude of questions. How can we understand the universe in which we find ourselves? How does the universe behave? What is the nature of reality? Where did all this come from? Did the universe need a creator? Most of us do not spend most of our time worrying about these questions but all most all of us worry about them some of the time. Traditionally these questions are for philosophy. But philosophy is dead. Philosophy has not kept up with modern developments. In science, particularly in physics, scientists have become the bearers of the torch of discovery in our quest of knowledge. The recent discoveries and theoretical advances lead mankind to a new picture of the universe and our place in it that is very different from the traditional one and different even from the picture we might have painted just a decade or two ago. In the history of science we have discovered a sequence of better and better theories or models from Plato to the classical physics theory of Newton to modern quantum theories. Will this sequence eventually reach an end point, an ultimate theory of universe that will include all the force and predict very observation we can make or will continue forever finding better theories but never one that can be improved upon? We do not yet have a definite answer of this question but we now have a candidate for the ultimate theory of everything, if indeed one exists, called M-theory. M-theory is the only model that has all the properties we think the final theory ought to have. M-theory is not a theory in usual sense. It is a whole family of different theories, each of which is good description of observation only in some range of physical situation. It is a bit like a map. Just like one cannot show the whole of the earth's surface on a single map. To map the entire earth one has to use a collection of maps, each of which covers a limited region. The map overlap each other end where they do, they show the same landscape. M-theory is similar to this. The different theories in the M-theory family may look very different but can all regarded as the aspects of the same underlying theory. They are the versions of the theory that are applicable only in limited ranges. The M-theory may offer the answers to the questions of creation. According to the M-theory ours is not only universe. Instead M-theory predicts that a great many universes were created out of nothing. Their creation does not require any intervention of any supernatural force or being. Rather the multiple universes arise naturally from the physical laws. They are a prediction of science. Each universe has many possible states and many possible histories in later times, that is, at times like the present long after their creation. Most of these states will be quite unlike the universe we observe and quite unsuitable for the existence of any form of life. Only a very few would allow the creatures like us to exist. Thus our presence selects out from this vast array only those universe that are compatible with our existence. Although we are puny and in significant on the scale of the cosmos, this makes us in a sense the lords of creation. To understand the universe at the deepest level, we need to know not only how the universe behaves but why, there is something rather than nothing, why do we exist? Why this particular set of laws and some others? This is the ultimate question of life, the universe and everything (Stephan Hawking Grand Design).

The nature-mankind friendly/centric philosophy is rooted in the concept of scientific determinism, which implies that there are no miracles or exceptions to the laws of nature. The universe is comprehensible because it is governed by scientific laws. But what are these laws or models. The first force to be described was gravitation, Newton's law of universal gravitation. The next aspect of the universe for which a law or model was discovered were the electric and magnetic forces. Our current idea of electricity and magnetism were developed over a period of about hundred years from the mid-eighteenth to mid- nineteenth century, when physicists in several countries made detailed experimental studies of electric and magnetic forces. Faraday's greatest intellectual innovation was the discovery of force field. Today we believe that all forces transmitted by the fields, so it is important concept in modern physics. For several years our understanding of electromagnetism remained stalled, amounting to no more than the knowledge of few empirical laws: The hint that the electricity and magnetism were closely related; the notion that they had some sort of connection to light and embryonic concept of fields. Then over a period of years in 1860 Scottish physicist James Clark Maxwell developed Faraday's thinking into a mathematical framework that explained the intimate and mysterious relation among electricity, magnetism and light.

Einstein's general theory of relativity reproduces special relativity when gravity is absent and it makes almost the same predictions as the Newton's law of universal gravitation in the weak gravitational environment of our solar system but not quite. In fact general relativity was not taken into account in GPS satellite navigation system, error in global positions would accumulate at a rate of about 10 km each day. However the real importance of general relativity is not its application in devices but rather it is very different model of the universe, which predicts new effects such as gravitational waves and Black Holes. The known forces of nature can be divided into four classes:

1. *Gravitational force:* It is the weakest of the four but it is the long range force and acts on everything in the universe as an attraction. This means that for large bodies the gravitational forces all add up and can dominate over all other forces.

2. *Electromagnetic forces:* This is also long range and is much stronger than gravity but it acts only on particle with an electrical charge being repulsive between the charges of the same sign and attractive between the charges of the opposite sign. This means the electric forces between large bodies cancel each other out but on the scale of atoms and molecules they dominate. The electromagnetic forces are responsible for all the chemistry and biology.

3. *Weak nuclear forces:* This cause radioactivity and plays a vital role in the formation of elements in stars and early universe.

4. *Strong nuclear forces:* The force holds together the protons and neutrons inside the nucleus of the atom. It also holds together the protons and neutrons themselves, which is

necessary because they are made up of still tinnier particles called quarks.

M-theory is the unified theory. The fact that we human beings who are mere collections of fundamental particles of nature have come this close to an understanding of the law governing us and our universe is a great triumph. But perhaps the true miracle is the abstract consideration of logic leads to a unique theory that predicts and describes the vast universe full of the amazing variety that we see. If the theory is confirmed by observation, it will be the successful conclusion of the search going back more than 3000 years. We will have found the grand design.

The various observations and results from WMAP (Wilkinson Microwave Anisotropy Probe, a spacecraft) team showed that there is no sufficient visible matter in the account for the apparent strength of gravitational forces within and between galaxies. There is a scientific assessment and understanding that universe today is 72% in the form of dark energy, 23% dark matter (that doesn't emit light and doesn't react with normal baryonic matter), 4% free hydrogen and helium, 0.5% stars, 0.3% neutrinos, and 0.03% heavy elements. The nature-mankind friendly global confederal model/system will fully utilize the global scientific-technological knowledge and mechanism to discover the total reality of universe (origin, creation, existence, motion, change and development) and search the life beyond the Planet Earth

NATURE-MANKIND FRIENDLY FUNDAMENTAL VISION

The nature-mankind friendly fundamental vision upholds that the nature and mankind are the two basic and most precious phenomenon of the human society. The existence, sustainability and rational advancement of the human society depend on the harmonious, balance and friendly relation between nature and mankind and within mankind. The nature-mankind friendly vision gives equal importance to both nature and mankind and always put the nature and mankind at the centre of every human activity in the world, discover life in the universe [beyond Planet Earth] and establish friendly relation with them. The nature and mankind are the two most important factors of human society in our Planet Earth. They constitute the capital of human society (the supreme phenomenon of the human society) and without them nothing has any value. The prevailing capital is only a medium of exchange operating in the form of money and commodity. The capitalist form of capital exists in the form of big concentration of money, large accumulation of commodities or stock market shares. The concept that society is based on money or capital is a wrong corporate capitalist concept. But in fact, the human society is based on two fundamental interconnected factors, nature and mankind. The change and development in human society takes place due to two-sided interaction between nature and mankind and within mankind. The two-sided interaction shows that the natural resources and the human resources constitute the highest valuable thing or the capital of the human society. The capitalist concept of the capital as the accumulation of money or wealth constitutes only medium of exchange.

NATURE-MANKIND FRIENDLY FUNDAMENTAL GOAL

The fundamental goal of the nature-mankind friendly global confederal model is to bring fundamental global social change by replacing the present existing unjust, irrational, anti nature and anti mankind global corporate capitalist system through democratic and peaceful global movement and to establish nature-mankind friendly global confederal system—the short-term way out and migration/settlement to safe place in the universe—the long-term way out.

FUNDAMENTAL GUIDING PRINCIPLES

Supremacy of Nature and Mankind

The nature-mankind friendly global confederal model firmly upholds the principal of supremacy of nature and mankind in human society. The nature and mankind are the two fundamental components and sustainable base, which plays basic role in rational advancement and survival of human society.

Scientific-Rational Naturalism cum Humanism

The principle of scientific-rational naturalism cum humanism is a new advance outlook of mankind. It is based on nature-mankind friendly/centric scientific-rational reasoning, present fundamental social reality of interdependent world (one world vision) and newly developing human community. The principle of scientific-rational naturalism cum humanism is the need of present global era to counter, tackle and solve the global threat of global warming/climate change and global dehumanization/human cum social inequality and external threat from universe. The nature-mankind friendly global confederal model sincerely uphold the principle of scientific-rational naturalism cum humanism, global unity of mankind, true human solidarity, transform and organize the divided mankind into a single global human community on the basis of nature-mankind friendly vision and scientific-rational naturalism cum humanism

Global People Govern Global Confederal Democracy

The word democracy is derived from the Greek word "demon and cratia". The word demon means people and cratia means power, i.e. people power. But throughout the history of national capitalism and global corporate capitalism, the present national democratic systems are regulating by the major national political parties, corporate capitalists and top bureaucrats. The people have only right to vote after four or five years. The real power is monopolized and centralized by the leading politicians of the major political parties, corporate capitalists, top bureaucrats, big landlords and mafias. The activists of the political parties and common voters are never consulted or involved for the selection of the election candidates, in major political decisions and in the formulation of policies. The candidates for election are always imposed by the high commands of the political parties in the world and the major political decisions and policies being formulated without the involvement of people and the activists of political parties. The new social reality of the world and mankind's fundamental challenge of global warming/climate change and global dehumanization/human cum social inequality have made the present party cum bureaucratic regulated national democratic systems anti nature-anti mankind, irrelevant and outdated.

The new social reality (a complex and broader global reality) created the historical need of active participation, involvement and full empowerment of the global people. The political parties of the world are the instruments of the old outdated capitalist national democratic systems. They emerged in national era. There was no existence of political parties in long human history before national era. The historical experience of human society tells us that they will not continue permanently. In fact, the mankind has entered into the global era which has made the national political parties of the world ineffective to carryout the agenda of 21st century, i.e. to establish the nature-mankind friendly global confederal future model/system of the mankind and safer settlement of mankind in other parts of universe beyond Planet Earth. There is a historical need to demonopolize and democratize the political powers of the political parties in the world and party regulated political mechanism in different countries by transferring the whole political power to the global people under fully democratized and confederal global political system.

The nature-mankind friendly global confederal model needs to launch global awareness and global mass education campaign on dangerous challenge of global warming/climate change and global dehumanization/human cum social inequality and in favour of future model of mankind, i.e. nature-mankind friendly global confederal model. The nature-mankind friendly global movement should play the role as a nature-mankind friendly guide of global people. It should truly ensure the active participation, involvement and dominant role of global people in people governance by transferring political power to the global people. The nature-mankind friendly global model upholds the new concept of global people governed-global confederal democracy. Under this new concept and model, the global people will be fully empowered and the global political power will be rationally distributed between global confederal govt and confederating units of the global confederal system on the basis of the people governance and confederal principle.

Human cum Social Fair Equality

The nature-mankind friendly global confederal model firmly upholds the principle of human cum social fair equality in nature-mankind friendly global human society. The nature-mankind friendly global confederal model should provide fundamental constitutional right of social-economic security to every deprived, unemployed and needy person of the world, should eliminate the irrational global human-social inequality and rationalise the irrational income differences, fully democratize the global political power system and global economy, i.e. to make the global people the masters of global political power system and global economy in real sense.

Global Transparency and Accountability

The nature-mankind friendly global confederal model truly upholds the principles of all round global transparency and global accountability in every global sphere of life from top to bottom.

NATURE-MANKIND FRIENDLY GLOBAL CONFEDERAL POLITICAL MODEL

Nature-Mankind Friendly Global Confederal State

The nature-mankind friendly global confederal state will be built on the basis of nature-mankind friendly global legislature, global executive and global judiciary. The nature-mankind friendly global confederal state/system will work without national borders (except national confederating demarcations) and without national wars. The global people will live and work unitedly as one human community with the orientation of rational advancement of nature and mankind and the fulfilment of genuine needs of nature and mankind, not human greed. The world will be democratically governed by the global people on the basis of nature-mankind friendly governess and global confederal model not by the power brokers and money holders.

Nature-Mankind Friendly Global Confederal Parliament

The nature-mankind friendly global confederal parliament will be constituted by the two houses, the house of mankind and the house of nature.

The House of Mankind

The house of mankind will be constituted by directly elected representatives from the existing confederating national units or reorganized confederating units of the global confederal system. The number of representatives of the confederating units (big or small) for the house of mankind will be decided in the global confederal constitution, which will be duly adopted through global referendum. The proceedings of the house of mankind will be chaired by the elected global secretary general or global chairman.

The House of Nature

The house of nature will be constituted by the representatives elected by the parliaments of the confederating-units of global confederal system. The representatives for the house of nature should be from natural scientists, global environment experts, global climatologists, nature-mankind friendly scientists, global experts of the natural-human resource development, global experts of disaster management and global experts of nature-mankind friendly economy or mankind regulated fully democratic economy (a economy owned, controlled and managed by the global people). The representation of confederal national units (big or small) in the house of nature will be equal and the number of representatives will also be decided in global confederal constitution, which will be duly adopted through global referendum. The proceedings of the house of nature will be chaired by the elected global chairman of the house of nature.

Nature-Mankind Friendly Global Confederal Missionary Government

The global confederal missionary govt will be strictly formed on the basis and with the tune of nature-mankind friendly philosophy, nature-mankind friendly fundamental vision and nature-mankind friendly global confederal model. The global confederal missionary govt will firmly implement and translate into practice the nature-mankind friendly global confederal model and also truly observe the principles of the global people empowerment (political, economical, cultural, diplomatic and security empowerment of global people), full global transparency (from global to bottom level), full honesty and global accountability. The global people governed global democratic system will be truly established. There will be very high level of global democracy, which will be actively run and regulated by the global people with nature-mankind friendly highest global morality. The global confederal system of missionary governance will be democratically regulated by the global people (with their full participation and involvement) along with their elected representatives and elected global confederal government. The global confederal missionary govt will have global confederal president or global confederal chief missionary along with global confederal missionary ministers. The global confederal president or global chief missionary will be directly elected by the global people or by the members of the global confederal parliament (both the houses of global parliament) through secret ballot. The global confederal missionary ministry will be duly approved by the global confederal parliament. The panel of the global confederal ministers will be submitted by the global confederal president or global chief missionary. The detail of the global missionary governance or government of nature-mankind friendly global confederal system will be formulated in the global confederal constitution. The main global headquarter of global confederal missionary government will be established with the approval of global parliament but the rotational headquarters will also be established in every continent.

Nature-Mankind Friendly Global Confederal Structure

The nature-mankind friendly global confederal system will organise and establish its systemic structure and global confederating units at four levels:

Nature-Mankind Friendly Global Confederal System

The nature-mankind friendly single global confederal system will be established in place of global corporate capitalist system which is operating through multiple nation-states. The present global era and new global realities demand to mobilize and organize the global people on the basis of nature-mankind friendly global confederal model for historical and revolutionary global change.

Nature-Mankind Friendly Continental Confederal System

The nature-mankind friendly continental confederal system will be established at Europe, Asia, Africa, North America, South America and Oceania.

Nature-Mankind Friendly Regional Confederal System

The nature-mankind friendly regional confederal systems will be established at (1) The South Asian Region, (2) The South-East Asian Region, (3) The Central Asian Region, (4) The Middle-East, (5) The East-African Region, (6) The South-African Region, (7) The Sub-Saharan African Region, (8) The Latin American Region, (9) The Caribbean Region, (10) The South American Region, (11) The Central American Region, (12) The North American Region, (13) The Region of Disintegrated USSR.

Nature-Mankind Friendly National Confederal Systems

The corporate capitalist systems of all the countries of the present corporate capitalist world will be restructured and remodelled with the tune of nature-mankind friendly global confederal model through peaceful and democratic global movement. The nature-mankind friendly global movement along with the global people will take the lead. The details of nature-mankind friendly global confederal structure and global confederating units will be formulated in the global confederal constitution and confederating constitutions, which will be approved through referendums (global, continental, regional and national referendums).

Nature-Mankind Friendly Global Confederal Subjects

The global confederal government will have the jurisdiction on the following subjects: (1) The nature-mankind friendly global human development commission or global department, (2) The nature-mankind friendly global sustainable scientific-technological development commission, (3) The global disaster management, (4) The nature-mankind friendly global peace force, (5) The nature-mankind friendly global economic development commission, (6) The global finance and taxation commission, (7) The nature-mankind friendly global medium of exchange and global financial commission, (8) The nature-mankind friendly global financial institution, (9) The nature-mankind friendly global exchange of goods and services commission, (10) Nature-mankind friendly global clean energy commission, (11) The nature-mankind friendly global health care commission, (12) The global peace and global human solidarity diplomacy commission, (13) The global environmentalist and scientific-rational humanist cultural commission, (14) The nature-mankind friendly global organic and co-operative agriculture commission, (15) The nature-mankind friendly global industrial manufacturing commission, (16) The nature-mankind friendly global organic food security commission, (17) the nature-mankind friendly global ocean development commission, (18) The nature-mankind friendly space and universe research and development commission, (19) The Nature-Mankind Friendly Global Forests and bio-diversity commission/department.

There will be global constitutional obligation for continental, regional and national confederating units of global confederal system to establish their social system with the tune of nature-mankind friendly global confederal model. But every confederating unit will enjoy full confederal autonomy and independence within the framework of nature-mankind friendly global confederal system. The details of division of subjects between global confederal government and confederating governments will be formulated in the global constitution, which will be approved through global referendum.

Global Fundamental Constitutional Rights

The global fundamental constitutional rights of the global people or the global mankind will be determined or based on global democratic principles of the global people governance and global people empowerment. The global people or global mankind will enjoy the fundamental constitutional rights as under: (1) The global fundamental constitutional right of sustainability of nature and mankind, (2) The global constitutional fundamental political right of global people or global mankind to govern the global confederal system, (3) The global constitutional fundamental political right of global people or global mankind to elect and recall the members of the global parliament, (4) The global constitutional fundamental political right of global people or global mankind to approve the global constitution of global confederal system through global referendum. The change in nature-mankind friendly global confederal model or global confederal systemic change or global constitutional change or any law related to global confederal system or global constitution can only be changed through the approval of global parliament and global referendum, (5) The global constitutional fundamental right to global fair equality in all spheres of human life (natural, social, political, economical, cultural, diplomatic, security and gender-equality). The global fair equality means maximum global equality in natural, social, political, cultural, diplomatic, security processes and global fair economic equality based on rationalization of income differences in global economy. To establish gender-equality (social, political, economical and cultural) between male and female in the whole world. The issue of gender equality is most important issue of the development of mankind. The gender equality is considered as the relation of the partnership between two equal partners. Historically it has been a relation of domination and subordination between man and woman under matriarchal system and patriarchal system. The nature-mankind friendly global confederal model upholds the empowerment (social, political, economical and cultural) of global women equal to men, (6) The global constitutional fundamental right to nature-mankind friendly free, compulsory and universal education up to higher level, (7) The global constitutional fundamental right to social-economic security with regard to sustainable climate, pure drinking water, right to work, right to house, organic nutritious food, global free health care and free from hunger (to poor, deprived, needy and unemployed persons), (8) The global constitutional fundamental right to full freedom of expression, individual cum collective freedom and liberty, (9) The global constitutional fundamental right to know and information for global people.

Fundamental Duties of Mankind

The whole mankind by enjoying the maximum global fundamental democratic rights (which makes the global people the real controller, regulator, administrator, governor and master of global human society) should also perform its duties and responsibilities individually and collectively as rational human species to sustain and rationally advance the human civilization. To save mankind as well as other bio-life on Planet Earth and discover life and conditions of life for the settlement of mankind in other planets and satellites of our solar system or other solar systems of the universe. The following duties and responsibilities are required from mankind individually and collectively as under:

To play the missionary role individually and collectively to save mankind and other bio-life on Planet Earth from the global threat of global warming/climate change and global dehumanization/human inequality and external threat from outer space beyond Planet Earth.

To firmly and truly uphold the nature-mankind friendly global confederal social model and actively contribute for its continuous updating with the tune of changes in nature and mankind and on the basis of new understanding about nature (universe) and sustainable-rational global human society.

The mankind should develop individually and collectively a most rational, fair, honest, super-moralist and super intelligent species of the universe by collectively (as a single human community and human family) developing organized universal intelligence and by equally promoting individual cum social dimensions of mankind. Every human being should truly involve and strive to unify and build the divided mankind into a single human community and human family.

Global Election System

There will be a global election commissioner headed by a global chief election commissioner. The global election commissioner will be constituted by the global parliament. The panel of global election commission will be submitted by the global government before global confederal parliament. The election system of present global corporate capitalist system operating through multiple nation-states is dominated by corporate capitalist money, unfair means and undemocratic methods. The nature-mankind friendly global confederal system upholds the election system of free, fair, value based, state funded and fully democratic. The voters of the world will have the right to select or give approval to candidates, prior to election of the global parliament to contest the election. The joint election campaign will be organized and funded by the global election commissioner. Nobody will be allowed to use the personal funds in global or other level elections. In the global election or the election of any confederating unit, the war mongering, the provocative (which incite ethnic conflict or human conflict or war), communal, religious fundamentalists, casteist and hatred propaganda will be strictly banned. The person who will be connected with war mongering organization, raciest organisations, narrow chauvinist organizations, religious fundamentalist organization, and casteist organization will not be allowed to contest election of global confederal parliament or any confederating unit. The global voters or the voters of any confederating unit will have the right to recall their elected representatives of global confederal parliament, if they loose their confidence on them.

The nature-mankind friendly global confederal model firmly uphold the principle of gender-equality and women political empowerment, for this a new democratic system of women political empowerment be established, under which there should be 50% provision in first phase to elect one female and one male from each constituency from grass root level institution to confederating parliament to global parliament (without harming the interest of men). This new democratic system should begin in 50% constituencies of every level of political Institutions in first phase and gradually increase to 100%.

GLOBAL JUDICIAL SYSTEM

The nature-mankind global confederal government will constitute the new global court of justice on the basis of new global jurisprudence and global confederal constitution with the approval of global parliament. The global disputes or disputes between confederating units will be settled by the global court of justice. The global confederal government will also constitute the global judicial commission with the approval of global parliament.

NATURE-MANKIND FRIENDLY GLOBAL CONFEDERAL SECURITY SYSTEM

Today mankind has reached into a global historical stage. In this new global historical stage all the nations of the world have become globalized and interdependent. The mankind is living in a global corporate capitalist system but the corporate capitalist system still operating mainly through nation-states. The nation-states are continuously loosing their validity and relevance due to growing globalization and existential threat of global warming/climate change and global dehumanization/human cum social inequality. Now the fundamental challenge and future of mankind has become common.

The global historical reality or global historical stage of human society and fundamental two-dimensional existential challenge of global warming/climate change and global dehumanization/human cum social inequality urgently and loudly demands the forging of sincere global human solidarity or true unity of global mankind and global unification or unification of the nations of the world (developed and developing nations) into a single nature-mankind friendly global confederal system. The global human solidarity and global structural unification needs the end of all the national disputes, disputes within nations, perspective of national supremacy, the approach of global domination, super-power approach and differences between rich industrialized developed nations and poor developing nations in the world. Today mankind needs total global peace, global human solidarity, global rational cooperation and global change from the military confrontation and war mongering politics to save nature-mankind global mission and global peace politics. The nature-mankind friendly global confederal system is the need of present global era and safer future of mankind. There will be no national wars and no national borders will exist in nature-mankind friendly global confederal system. The nature-mankind friendly global confederal model firmly uphold to dismantle all the nuclear and mass destruction weapons and the national defence and military systems of all the countries of the world. The military and defence sector is the biggest global sector of wasteful and unproductive expenditure. The global confederal government will take the exclusive responsibility of global security of global people. The nature-mankind friendly global confederal system will build a "nature-mankind friendly global peace force". The people of dissolved national armies will be engaged in (1) The nature-mankind friendly global peace force (2) In global disaster management (3) In protection and promotion of global forest cover, global wildlife, bio-diversity, oceans and in other alternative suitable places according to their ability. The trillions of amount of money saved from national defence sectors of different countries will be spent on the conservation, promotion and development of nature, global environment repair, oceans protection and study, protection and promotion of global forest cover, global wildlife, global bio-diversity and development of mankind, poverty elimination, elimination of inequality and dehumanization, the development of nature-mankind friendly sustainable and clean technology and the development of nature-mankind friendly clean and sustainable energy particularly in place of fossil fuels, thermal energy and risky nuclear energy. The big gap between rich developed nations and poor developing nations will be reduced and rationalized.

NATURE-MANKIND FRIENDLY GLOBAL CONFEDERAL ECONOMIC MODEL

The nature-mankind friendly global confederal model/system consider that the market regulated global corporate capitalist model or unsustainable global corporate capitalist development model/system which is existing and operating today throughout the world, is facing serious global crisis and heading towards collapse and the state regulated socialist-communist development model/system (which has failed and discarded almost everywhere in the world) has proved one sided and anti nature-anti mankind. Both the economic or development models are the creator of fundamental existential challenge of global warming/climate change and global dehumanization/human cum social inequality (global poverty, unemployment, hunger, etc.) and have become ineffective and irrelevant. The mankind cannot solve the basic global problems (created by global corporate capitalist system) of global warming/climate change and global dehumanization/human cum social inequality, global poverty, global unemployment, global hunger, global rising prices and global cyclic economic recession within the framework of global corporate capitalist system or global corporate capitalist economic model (unsustainable development model). The nature-mankind friendly global model firmly uphold and advocate the replacement of the global corporate capitalist system or global corporate capitalist economic model (unsustainable development model) and to establish the nature-mankind friendly global confederal economic model (development model) in the light of nature-mankind friendly global confederal model, its philosophy, fundamental vision and fundamental goal.

Fundamental Features of Nature-Mankind Friendly Global Confederal Economic Model

Nature and Mankind two Basic Factors of Economic Development

The nature-mankind friendly global confederal economic model upholds that the nature and mankind are the two most precious and equally important phenomena in the human society and constitute the global capital. The sustainable development of natural resources and human resources or nature-mankind friendly balance development is the real indicators of the social development, progress and prosperity, instead of one-sidedly focusing on growth rate without caring nature and mankind. The nature and mankind are the two basic factors of sustainable development. The nature-mankind friendly development model accords top priority to nature and mankind during global development or economic development process and firmly adopts the nature-mankind friendly global economic development model.

Global Confederal Economic Model

The nature-mankind friendly global confederal model firmly committed to build nature-mankind friendly single global confederal economic development system, in the light of nature-mankind friendly global confederal model. The corporate capitalist interdependent national economies of all the countries of the world will be remodelled on the basis of the nature-mankind friendly global confederal economic model and will be made the confederating economic units of global confederal economic system. The global confederal government with the approval of global confederal parliament will setup the nature-mankind friendly global sustainable development commission to act as a global confederal authority for the conservation and promotion of global natural resources and the rational cum equitable development of global human resources. The nature-mankind friendly global sustainable economic development commission will work for the rational advancement of nature and mankind. The global sustainable economic development commission will propose the enactment of nature-mankind friendly global laws and the proposals of global advancement and global sustainable institutions.

Economic Empowerment of Global People

The economic empowerment of global people will be based and established by ensuring constitutional right of democratic ownership of global people and democratic management of nature-mankind friendly global confederal economy by global people or global mankind, constitutional fundamental right of social-economic security and fair global economic equality.

People Democratic Ownership of Nature-Mankind Friendly Global Confederal Economy

The nature-mankind friendly global economy will be owned, democratically managed and regulated by the global people or global mankind. After dismantling and replacing the global corporate capitalist economy and abolishing of global corporate capitalist ownership (including big landlord ownership) the global people or the entire mankind will be made the owner of the global confederal economy. The corporate capitalist private sector or the corporate capitalist public sector will be dissolved and nature-mankind friendly global people regulated new democratic global confederal economic system will be built. The new democratic and global confederal economic system will not be regulated through global corporate capitalist market perspective or global corporate capitalist market mechanism but it will be regulated through nature-mankind friendly perspective and nature-mankind friendly mechanism. The nature-mankind friendly global confederal model out rightly reject the market centric and state centric perspective or mix-up of state and market perspective, mechanism and regulation. The people regulated global confederal democratic economic system will give 70% ownership rights or democratic share holdings to the global people (i.e. workers, employees, poor, deprived, needy, unemployed, the people of lower middle class, scientists, experts, engineers, doctors, etc.) in nature-mankind friendly industrial manufacturing sector, nature-mankind friendly global organic-cooperative agriculture sector, nature-mankind friendly service sector and nature-mankind friendly global infrastructure of global confederal economy. The 30% ownership rights or democratic share holdings will be retained by the global confederal government or confederating governments. In the small scale level in every sector 100% ownership rights will be given to the people and in some bigger units of every sector 100% holding will be retained by the state powers. This is a tentative people ownership economic model. But the democratic ownership of people, global confederal state and confederating states on global economy will be further rationalized and made more realistic with the tune of continuous and periodical deep study, deep investigation and global experience.

Global Fair Economic Equality

The nature-mankind friendly global confederal model will ensure the global fundamental constitutional right of fair economic equality in the form of social-economic security (with regards sustainable environment-climate, pure drinking water, organic nutritious food, nature-mankind friendly housing, free health care, right to work and free, compulsory and universal global education up to higher level) to every poor, deprived, needy, unemployed and lower middle class people of the world. To rationalize the irrational global income differences. To eliminate and end the global gender inequality, rich and poor nations gap, urban and rural gap, have and have not gap by establishing fair global equality in every sphere particularly in global economy.

Global People Regulated Democratic Management

The global corporate capitalist model practicing the corporate capitalist monopoly management and the state regulated socialist-communist model stands for bureaucratic monopoly management. But the nature-mankind friendly global confederal economic model upholds the establishment of people regulated democratic management. The democratic management of different democratic sectors (global industrial manufacturing sector, global organic-cooperative agriculture sector, nature-mankind friendly service sector and nature-mankind global infrastructural sector) of global confederal economy or economies of confederating units will be constituted by 70% representation of global people or people of confederating units in the "management boards" (i.e. workers, employees, poor, deprived, needy, unemployed, the people of lower middle class, scientists, experts, engineers, doctors, etc.) and 30% representation from global confederal government or confederating governments.

Nature-Mankind Friendly Global Organic-Cooperative Democratic Agriculture System

The agriculture is an oldest and biggest sector of the world economy. The agriculture sector is a lifeline of global mankind. Today mankind is facing a serious global agriculture crisis. The chemical technology based (chemical fertilizers, pesticides, insecticides, herbicides, etc.) inorganic global agriculture system (started in the name of green revolution) has created a serious global agriculture crisis. The global agriculture crisis is reflecting in the form of damaged pest resistance of agriculture soil and killing of crop friendly micro-organism, production of poisonous inorganic food (which is the cause of so many deadly diseases), the emission of green house gases (methane and nitrous oxide), which are inducing along with other green house gases the global warming/climate change and pollution of underground water. The lack of priority and backwardness of agriculture sector due to the wrong policies of the different national governments of the world, continuous fragmentation of agriculture land, increasing agriculture cost of production, lack of appropriate prices of agriculture products and unfair global agriculture trade, corporatizing of agriculture sector, construction of special economic zones, construction of commercial complexes and shopping malls in agriculture lands, stress of growing global population on agriculture land, serious impact of global warming-climate change on agriculture sector, have seriously affected the global agriculture system in general and impoverishment of the poor and medium farmers in particular.

The newly introduced solution of genetically modified technology of agriculture (after the realization of negative effects of chemical based inorganic farming) and the corporate capitalist proposal of contract farming is not the solution of global agriculture crisis. The time demands fundamental global change in global agriculture system. The nature-mankind friendly floban confederal model/system firmly uphold to change the present global chemical technology based unsustainable anti nature-anti mankind global agriculture system and to establish nature-mankind friendly global organic-cooperative [integrated multi-purpose people farming] democratic agriculture system. The system of global organic-cooperative [integrated multi-purpose people farming] democratic farming will be different from the pattern of socialist collective farming which had deprived the farmers from ownership rights and made the state ownership. Under the new system of global organic-cooperative farming the fragmented agriculture land will be consolidated into big integrated multi-purpose organic-cooperative farms. The 70% ownership rights in the agriculture farms will remain with the farmers who will contribute land to the integrated organic-cooperative farms and 30% ownership rights will be given to the confederating governments of the world which will provide the required funds for the development of integrated multi-purpose organic-cooperative farms. There will be democratic management in organic-cooperative integrated farms through the 70% representation of concerned farmers and 30% from the global confederal government or confederating governments in the management board. To accord nature-mankind friendly agro-industrial status to the organic-cooperative farms. The details of global organic-cooperative farming will be formulated separately in global agriculture model/policy.

Nature-Mankind Friendly Global Industrial Manufacturing Democratic Sector

The nature-mankind friendly global confederal model firmly committed to fully restructure and remodel the global corporate capitalist industrial manufacturing sector and to build a new n..ture-mankind friendly global industrial manufacturing democratic sector on the basis of nature-mankind friendly global confederal model. In this new nature-mankind friendly global industrial manufacturing democratic sector the 70% ownership rights or democratic share holdings will be given to global people (i.e. workers, employees, unemployed, poor, deprived, scientists, experts, engineers, doctors and general people of the world) and 30% ownership rights or democratic share holdings will remain with confederal government or confederating governments. The management of nature-mankind friendly global industrial manufacturing democratic sector will be made fully democratic. The 70% representation for 'management board' will be given to the world people (i.e. workers, employees, poor, deprived, needy, unemployed, scientists, experts, engineers, doctors, the people of lower middle class and the general people of the world) and 30% representation will be from global confederal government or confederating governments. The polluting technologies, energy and raw materials from this sector will be totally replaced and nature-mankind friendly sustainable and new clean technology, clean energy and clean raw materials will be introduced.

Nature-Mankind Friendly Global Service Sector

The global corporate capitalist service sector will be restructured and remodelled on the basis of nature-mankind friendly global confederal model, especially the transport sector and information technology sector.

In nature-mankind friendly democratic global service sector the 70% ownership rights or democratic share holdings will be given to global people (i.e. workers, employees, unemployed, poor, deprived, scientists, experts, engineers, doctors and general people of the world) and 30% ownership rights or democratic share holdings will remain with confederal government or confederating governments. The management of nature-mankind friendly democratic global service sector will be made fully democratic. The 70% representation for 'Management Board' will be given to the world people (i.e. workers, employees, poor, deprived, needy, unemployed, scientists, experts, engineers, doctors, the people of lower middle class and the general people of the world) and 30% representation will be from global confederal government or confederating governments. The polluting technologies, energy and raw materials from this sector will be totally replaced and nature-mankind friendly sustainable and new clean technology, clean energy and clean raw materials will be introduced.

Nature-Mankind Friendly Global Infrastructure Sector

The global corporate capitalist infrastructure sector damaged the nature and mankind immensely. The nature-mankind friendly global confederal model considers the present corporate capitalist infrastructure sector as an anti nature-anti mankind sector. It will be fully replaced or remodelled, particularly, the global energy sector through global clean energy sector with the tune of nature-mankind friendly global confederal model. The special focus will be given to the use of nature-mankind friendly global clean energy, clean raw materials and clean Infrastructure technology. In nature-mankind friendly democratic global infrastructure sector the 70% ownership rights or democratic share holdings will be given to global people (i.e. workers, employees, unemployed, poor, deprived, scientists, experts, engineers, doctors and general people of the world) and 30% ownership rights or democratic share holdings will remain with confederal government or confederating governments. The management of nature-mankind friendly democratic global infrastructure sector will be made fully democratic. The 70% representation for 'management board' will be given to the world people (i.e. workers, employees, poor, deprived, needy, unemployed, scientists, experts, engineers, doctors, the people of lower middle class and the general people of the world) and 30% representation will be from global confederal government or confederating governments.

Nature-Mankind Friendly Sustainable Productivity

The nature-mankind friendly global confederal economic model will rationally use the global natural resources and global human resources. The process of global production will be developed and carry on with the strict application of fundamental vision of always focus on nature-mankind friendly development or human activity which always and all times maintain nature-mankind friendly balance and relation.

Rational Balance of Individual and Social Interest

The nature-mankind friendly global confederal model firmly rejects the individual centric approach of corporate capitalist model propounded by capitalist philosopher Adam Smith that man is selfish by nature and workers centric or general centric or social centric approach of Karl Marx or communist model. The nature-mankind friendly global confederal model strongly upholds that man/mankind is bio-social, having two fold nature, individual cum social. The human science and rationality demands that we should always maintain the rational balance between individual and social interests. The nature-mankind friendly global confederal model will firmly observe and maintain the rational balance between social and individual interests.

Nature-Mankind Friendly Global Planning Commission

The global confederal government will constitute the global confederal planning and sustainable development commission with the approval of global confederal parliament. The nature-mankind friendly planning and development commission will formulate the short term and long term nature-mankind friendly fully democratic sustainable development perspective planning contrary to corporate capitalist unsustainable, one-sided, anti-nature and anti-mankind, growth and profit friendly centralized planning.

Global Confederal Medium of Exchange

The global confederal government will introduce, with the approval of global people through global referendum, a new system of medium of exchange which will be guided and regulated through the nature-mankind friendly fundamental perspective and global mechanism.

Nature-Mankind Friendly Global Exchange of Goods and Services Commission

The global confederal government will constitute the global exchange of goods and services commission with the approval of global confederal parliament. The global exchange of goods and services will be regulated with the nature-mankind friendly perspective not with the corporate capitalist profit maximization motive or perspective.

Nature-Mankind Friendly Global Confederal Taxation Commission

The global confederal government with the approval of global confederal parliament constitutes the global confederal taxation commission. This commission will regulate and manage the global taxation system.

Global Confederal Financial Commission

The global confederal government with the approval of global confederal parliament will constitute the global confederal financial commission and global financial institution.

NATURE-MANKIND FRIENDLY GLOBAL DIPLOMACY

The global corporate capitalist diplomacy is based on the perspective of national supremacy and global corporate capitalist aim of profit and wealth maximization, controlling and using of global natural resources and cheap human resources. In fact, the global corporate capitalist diplomacy is based on the market perspective and corporate capitalist global design for the sole purpose of growth maximization, profit maximization and wealth maximization. The nature-mankind friendly global confederal model will fundamentally change the narrow nationalistic and market oriented, outdated global corporate capitalist diplomacy. The nature-mankind friendly global confederal system will adopt a new nature-mankind friendly global diplomacy. The new global diplomacy stands for:

1. To save nature (especially the environment of Planet Earth) and mankind from the danger of global warming/climate change and global dehumanization/human-inequality

2. To build universal brotherhood on the basis of nature-mankind friendly outlook
3. To uphold the principle of resolution of all the disputes of the world peacefully and democratically on the basis of just and rational compromising solution in the light of nature-mankind friendly outlook.
4. The nature-mankind friendly global confederal model considers that the J&K problem is a historical-political problem.

The complex J&K problem is the outcome of communal national vision, communal partition of India in 1947, tribal attack of Pakistan on J&K in 1947, [then the princely J&K state was divided between India and Pakistan with the demarcation of cease-fire line and now LOC, which is still continuing] the continuous Indo-Pak hot war-cold war [which had immensely damaged both India and Pakistan] due to communal mindset and confrontationist politics since 1947 and the emergence of Kashmiri muslim militancy and separatist movement from 1989, rising voice of Jammuites against the discrimination and in favour of their political rights, rising movement of Ladhakhi people for Union Territory status, Panan Kashmir movement of migrated Kashmiri Pundits, emerging demand of autonomous status for Poonch-Rajouri Pathwari ethno-region and Chenab valley ethno-region. Now the J&K has become a dangerous global nuclear-flashpoint of the world. The historical genesis tell us that, in the very inception of J&K problem, initially it had become a bilateral problem between India and Pakistan but due to the emergence of different movements in the J&K state it has become a trilateral problem among India, Pakistan and the people of J&K. The different bilateral accords between India and Pakistan and between Union Govt of India and J&K [Indian administered J&K part] state govt in the past have not proved effective because they were not in the tune of trilateral nature of J&K problem. The latest ongoing mass uprising and mass movement in Kashmir valley from June 2010 [in which more than 110 people, mostly youths killed] forced us to think about the just, rational and permanent settlement of J&K problem. But the biggest stumbling block before the settlement of J&K problem is the fundamentalist communal mindsets, confused narrow hard lines and unreal options [contrary to the trilateral nature of J&K problem] of different forces involved in the J&K problem and lack of consensus among the different forces and parties. The historical experience of J&K problem, the emerging reality of interdependent world [globalized world], the global challenge of global warming/climate change and global dehumanization/human inequality and social rationality demands from all the forces/parties that to initiate immediately the meaningful trilateral/tripartite dialogue process in the form of tripartite conference among India, Pakistan and the representatives of J&K people [representatives from the Kashmir region, Jammuites from Jammu region, the Kashmiri pundits, the West Pak refugees of 1947, Poonch-Rajouri Region, Chenab valley region, Leh and Kargil regions, Poonch-Bhimber-Mirpur region in Pakistan administered J&K, Baltistan-Gilgit or northern area in Pakistan administered J&K, including all the militant groups and separatist groups.] In the first phase of trilateral dialogue process all the forces/parties should strive honestly and sincerely to hammer out a realistic and broader consensus in the tune of trilateral nature of J&K problem. In the consensus process primarily, to declare the compromised trilateral cease-fire, to shun the path of violence and armed suppression and change the unilateral and bilateral approaches. During the trilateral dialogue and consensus process all the options and solutions of J&K problem should be discussed thoroughly without any pre condition. This will create a positive atmosphere of confidence building, free and frank discussion, mutual understanding and rationalize the unreal and diehard approaches of the parties involved in the J&K problem. The nature-mankind friendly global confederal model propose a compromising trilateral solution of J&K trilateral problem for the discussion and consideration to all the forces/parties as under—to establish Indo-Pak joint administration over entire Jammu-Kashmir state (both the divided parts of J&K) with regard joint security, joint external affairs, joint currency, common initiative to save nature-mankind from the threat of global warming/climate change and global dehumanization/ human inequality, common sustainable trilateral develop-mental projects and common clean energy projects. The rest of J&K state should be made fully autonomous federal state. The autonomous federal J&K state will have its own federal assembly. Under autonomous federal J&K state all the ethno-regional identities should be accorded the regional autonomous status in the form of regional council/assembly, i.e. (1) The Kashmiri regional council/assembly, (2) The Jammu regional council/assembly, (3) The Poonch-Rajouri Pathwari regional council/assembly, (4) The Chenab valley regional council/assembly, (5) The Leh-Kargil council/ assembly, (6) The Poonch-Bhimber, Mirpur council/assembly (Pathwari speaking region in Pakistan administered J&K), (7) The Baltistan-Gilgit, i.e. northern areas council/assembly, (8) The ethnic Kashmiri Pundits be settled in Kashmir after their approval, (9) The west Pakistani refugees should be made permanent resident of J&K state. The divided J&K state should be united into one united autonomous federal state. The formula of compromising solution of J&K trilateral problem in real sense addresses reconciles and harmonizes the genuine interests of India, Pakistan and J&K state and conforms to the ground reality and basic nature of J&K problem.

This trilateral compromising solution will unify the divided J&K state. The united J&K state will be a strong bridge between India and Pakistan. It will also help to solve the outstanding Indo-Pak disputes, will undo 1947's communal partition of India and Pakistan and lay sound foundation for the unification of South Asian region into nature-mankind friendly SAARC confederation. It will also promote the permanent South Asian regional and global peace, stability, progress and will promote the solidarity of mankind globally. Similarly all the disputes of the world should be solved especially the Middle East dispute of Palestine-Israel, Indo-China border dispute, Sri Lankan Tamil-Sinhalese dispute, Korean peninsula dispute between North Korea and South Korea, withdrawal of NATO forces from Iraq and Afghanistan, closing of US military bases throughout the world, dismantling of all nuclear weapons under the supervision of UN General Assembly or Global Confederal System. The restructuring and democratization of UNO by abolishing the veto power of five permanent members is the need of present global era. This will be a great leap in the history of Asia and the world.

The nature-mankind friendly global confederal model sincerely believes that the Indo-Pak historical dispute since 1947 is the biggest stumbling block before the development and advancement of India and Pakistan. The Indo-Pak dispute continuously creating Indo-Pak confrontation and hot war-cold war, which had immensely damaged to both India and Pakistan. Now India and Pakistan have become nuclear States and further Indo-Pak military conflict/war will be very dangerous and destructive for India and Pakistan in particular and mankind in general. The nature-mankind friendly global confederal model uphold that today we are living in a Interdependent globalized world, where the vision of narrow nationalism, intra-national conflicts and national wars have become irrational, irrelevant and contrary to the fundamental reality of interdependent globalized era, which demands the building of nature-mankind friendly single global confederal system. The nature-mankind friendly global confederal model appeals to both India and Pakistan to resolve peacefully and democratically all the Indo-Pak disputes/problems [J&K problem, Sir Creek dispute, Sichen glacier dispute, Wullar-Barrage dispute, water dispute, etc.] on rational cum give and take basis and take sincere pledge for Indo-Pak peace and strategic co-operation on environmental, political, economical, social, cultural, diplomatic and security spheres and jointly work for the building of nature-mankind friendly south Asian [SAARC] confederation.

The Indo-Pak and Indo-China disputes/conflicts are the biggest irritants, security threat and impediment before Asian development on the one hand and on the other hand the Asian continent is becoming global focus and engine of global economy. China and India are emerging as biggest corporate capitalist players of the global corporate capitalist system. China and India are the emerging economic powers of the world, so the Indo-China political and military confrontation will be very counter productive and disastrous. The nature-mankind friendly global confederal model firmly upholds that the Indo-China border/territory dispute should be solved peacefully on the basis of rational compromising solution without any delay. India and China should adopt the approach of mutual and balance of interests and sincerely take care about each other's interests and sensitive concerns. They should firmly and sincerely follow the path of strategic partnership and co-operation in particular and global co-operation in general, instead of destructive confrontation. The nature-mankind friendly global confederal model appeals that India, China, Japan, Russia and Pakistan should form a strategic group of Asia to solve their disputes and all disputes of Asia permanently. To work jointly for the building of nature-mankind friendly Asian confederation.

The middle east dispute between Israel and Palestine [including Arab countries] is a very old dispute/conflict and has become a dangerous bloody spot of the world. The middle-east dispute should be solved peacefully through dialogue among the Israel, Palestine and Arab countries involved in the dispute, [including Hamas and PLO] on the basis of a just and rational compromising solution, in which to recognise both Israel and Palestine states with long-term perspective of nature-mankind friendly middle east confederation.

The conflict/dispute of North and South Korea in Korean Peninsula is also a terrible and sensitive spot of the world. The Korean Peninsula dispute of North Korea and South Korea should be solved peacefully with the co-operation of China and US with the strategic perspective of nature-mankind friendly North-South Korean confederation.

The peaceful resolution of disputes of the world will help to promote real global peace, global brotherhood and global human solidarity, global unification for the advancement of next global historical stage and enable the mankind to tackle effectively the fundamental global challenge of global warming/climate change and global dehumanization/human cum social inequality and will facilitate to build new nature-mankind friendly global confederal system. The nature-mankind friendly global diplomacy will strive to change the outdated thinking and mindset of global people and prepare them for new global historical change. The nature-mankind friendly global confederal government will constitute the global diplomacy commission with the approval of global parliament.

NATURE-MANKIND FRIENDLY GLOBAL VALUE SYSTEM AND GLOBAL CULTURE

The nature-mankind friendly global confederal model need to launch global democratic campaign to fundamentally change the traditional pre-national fanatic and corporate capitalist, selfish, greedy, double standard, inhuman, violent, criminal, unequal and discriminatory culture. The nature-mankind friendly global confederal system will adopt the nature-mankind friendly new global value system and rational humanist-environmentalist global culture. The nature-mankind friendly sustainable and rational human way of life or global rational lifestyle is the need of human civilization and nature-mankind friendly global confederal model. The global challenges demand from the mankind, to adopt new global culture of global brotherhood and rational human solidarity on the basis of nature-mankind friendly perspective. The challenges and future of mankind or global people has become common and global. The organization of rational global human community is the historical necessity of global era. The nature-mankind friendly global confederal system will sincerely observe the rational value system of, the conformity of theory and practice, the conformity of means and ends, the maintenance of rational balance of individual and social interests in accordance with the bio-social nature of mankind and to shun the animal instinct of violence and selfishness. The nature-mankind friendly global confederal model upholds the new rational system of marriage on the basis of scientific principles cum gender equality and rejects the unscientific marriage ceremonies. The new global approach about the death and disposal of human dead bodies will be adopted. The details of new marriage codes and new death codes will be separately formulated.

The nature-mankind friendly global education system will be established on the basis of nature-mankind friendly global confederal model. There will be nature-mankind friendly universal, compulsory and free global education up to the higher level for the global people. The nature-mankind friendly global confederal system will give fundamental right of free health and medical care to global people. The high thinking, high morality, high value system, rational humanist behaviour of love and affection, peaceful co-existence, individual cum collective rational development and live as a single broader "human family" in a emerging global human society or global village is the loud call of single human civilization or the whole mankind of the Planet Earth.

NATURE-MANKIND FRIENDLY GLOBAL CONFEDERAL CONSTITUTION

The nature-mankind friendly global confederal system will be established on the basis of nature-mankind confederal global model, which will be reflected in global confederal constitution. The nature-mankind friendly global confederal constitution will be drafted by the global confederal parliament and will be duly approved by the global people through global referendum. The amendment in nature-mankind friendly global confederal constitution will be done only through global referendum.

8

Implementation of Save Nature-Mankind Global Agenda: Nature-Mankind Friendly Global Confederal Model

The Nature-Mankind Friendly Global Confederal Model is firmly and truly committed as a missionary global movement to launch global mass movement on global agenda of save nature and mankind from the fundamental global challenge of global warming/climate change and global dehumanization/ human cum social-inequality. The fundamental goal of Nature-Mankind Friendly Global Confederal Model is to fundamentally change and replace the present global corporate capitalist system through global democratic movement and establish nature-mankind friendly new global confederal system. The implementation process of two-fold global agenda of save Planet Earth and mankind from the global warming/climate change (including outer danger from the space) and global inequality/dehumanization and the implementation of Nature-Mankind Friendly Global Confederal Model is based on multi-interconnected aspects.

THEORETICAL AND AWARENESS ASPECT

The historical experience of social changes in the history of human society shows that no fundamental social change can take place if the concern people remain unaware about the historical need of such fundamental change. The fundamental social change requires moulding of the old traditional and corporate capitalist mindset of the global people. The transforming of basic agenda, the fundamental goal or the Nature-Mankind Friendly Global Confederal Model of nature-mankind friendly global movement into reality, depends on the changing of mindset of the global people. The nature-mankind friendly global movement will launch global mass education and awareness campaign with full motivation and dedication in order to change the traditional ideology and thinking of the global people The inculcation of higher education and deep understanding in the minds of missionaries and activists of the nature-mankind friendly global movement about Nature-Mankind Friendly Global Confederal Model is the dire necessity and focus of global mass awareness and mass education campaign. For this purpose special educational and training classes of missionaries and activists should be organized and conducted by the experts of Nature-Mankind Friendly Global Confederal Model and the experts of global warming/climate change and global dehumanization/human cum social inequality. The global mass education and mass awareness campaign will be conducted in an organized manner by the missionaries and activists of the nature-mankind friendly global movement from grass root level to global level, through study circles, awareness conferences and public meetings. The maximum possible involvement of people in awareness conferences, study circles and public meetings will be the priority of nature-mankind friendly global movement. The awareness campaign will also be launch through books, literature, leaflets, posters, dramas, films etc. The global mass awareness campaign will be linked with the mass movement about mass problems and mass issues at every level.

THE ASPECT OF GLOBAL MOVEMENT

The global people should launch well-organized democratic mass movement under the banner of nature-mankind friendly global platform, against global warming/climate change and global dehumanization/human cum social inequality along with mass movement on mass problems (both individual and common problems) and mass issues and against global corporate capitalist system in every country of the world from the grass root level to global level. The motivators and missionaries of the Nature-Mankind Friendly Global Confederal Model should also actively participate in the elections of present system all over the world as a part of global movement on the basis of its own people governed confederal democratic norms. The maximum participation and involvement of the global people in democratic mass movement under the banner of nature-mankind friendly global platform should always remain first priority. The nature-mankind friendly global platform should fully mobilize the global people to replace the global corporate capitalist system through global mass movement and to establish new Nature-Mankind Friendly Global Confederal System. The nature-mankind friendly global movement should adopt the democratic and peaceful means and methods during global mass movement. Every action or movement at any level and at any place against global warming/climate change and global dehumanization/human cum social inequality and global corporate capitalist system or in favour of Nature-Mankind Friendly Global Confederal Model will be considered as the integral part of nature-mankind friendly global mass movement.

There should be conformity in our theory and practice and our means and ends. Every guiding centre (missionary social unit) of nature-mankind friendly global platform should prepare its concrete and comprehensive action plan with the full involvement, participation, consultation and approval of concerned people. Every guiding centre (missionary unit) will take conscious initiative and action on the basis of the approved action plan instead of spontaneous response to the people. Every missionary and activist will adopt the approach of active initiator, instead of passive respondent to the people's initiative individually or collectively.

ORGANIZATIONAL ASPECT

The nature- mankind friendly global mass movement needs a nature-mankind friendly global mass organization. After awareness and mobilization we should involve and organize the global people or global mankind into nature-mankind friendly global movement (global organization of new model) on the basis of its fundamental global guiding model and organizational rules.

PROPAGANDA ASPECT

The Nature-Mankind Friendly Global Confederal Model will fully use the existing electronic and print media of the world, film industry of the world and the global system of internet or other sources of propaganda at its disposal. Every missionary and activist member of Nature-Mankind Friendly Global Confederal Model should launch consciously and seriously the regular mass propaganda, individually and collectively. The nature-mankind friendly global movement should prepare propaganda material and distribute it among the masses.

THE ASPECT OF FUNDS

The nature-mankind friendly global movement should collect the funds from the global people. Every supporter, activist and missionary of nature-mankind friendly global model should be missionary duty bound to contribute funds according to his or her financial capacity and also will take the responsibility to mobilize the masses to contribute to nature-mankind friendly global movement. The accounts of income and expenditure should be properly maintained and duly audited and the annual balance sheets of every level of nature-mankind friendly global movement should be presented before the people

FUNCTIONING ASPECT

The Nature-Mankind Friendly Global Confederal Model demands that our functioning style and methods of work should be fully democratic, rational and nature-mankind friendly. The Nature-Mankind Friendly Global Confederal Model will adopt organized and modernized system of functioning from global to grass root level. The Nature-Mankind Friendly Global Confederal Model will continuously update and modernize its system of functioning with the use of modern and latest skills. The maximum participation and involvement of the global people in global mass movement under the banner of nature-mankind friendly global movement should always remain our first priority and to adopt the democratic and peaceful methods during mass movement. The global people should build the nature-mankind friendly global movement under global platform and prepare concrete and comprehensive action plan with the full participation, involvement, consultation and approval of concerned people and should take conscious initiative and action on the basis of approved action plan. To adopt the principle of humble dialogue and discussion as a method to take decisions, to convince and motivate others and to adopt the principle of consensus as the main method of decision making by fully involving the participants in the discussion. To launch peaceful and democratic global mass movement against global threat of global warming/climate change and global dehumanization/human inequality, against global corporate capitalist system, on mass problems and mass issues and for the establishment of alternative Nature-Mankind Friendly Global Confederal System–in the form of submitting memorandum to the administration and governments for the solution of mass problems, mass meetings, holding seminars, debates, workshops, conducting mass education and training classes, organizing cultural programme, mass dharnas, long march, mass processions, agitations and demonstrations. To build strong relations and solidarity with the people movements of the world who are fighting against the global warming/climate change, global dehumanization/human inequality, human rights violation and global mass issues and to strive to unite the environmental mass movements and other people movements under the single global banner against the threat of global warming/climate change and global dehumanization/human inequality and to replace the global corporate capitalist system in order to establish Nature-Mankind Friendly Global Confederal System. The nature-mankind friendly global movement should establish nature-mankind friendly global research centre and nature-mankind friendly research centre in every country. The nature-mankind friendly global movement should also establish nature-mankind friendly missionary centres throughout the world, at every level and every place.

DELHI DECLARATION

The Indian government as the host of cop 8 set as a principal objective for the adoption of a Delhi Declaration, a broad political statement meant to signify the meeting's success. The initial draft circulated by the Indian chair of the conference reflected a strong perspective of developing countries, emphasizing the issues of sustainable development, adaptation and implementation of commitments by the developed countries under the UNFCCC. The draft was silent on the question of steps beyond Kyoto's first commitment period (2008-2012), prompting strong objections from the European Union and some other developed countries. The United States was largely contented with the Indian draft. The G-77 representing developing countries, called for a stronger emphasis on financial assistance and on the adverse economic effects on developing countries due to measures taken to reduce green house gas emissions.

The Delhi ministerial declaration on climate change and sustainable development makes no reference to future steps to further elaborate the climate regime. It largely underscores principles established in the UNFCCC and themes adopted at the World Summit on sustainable development earlier in Johannesburg. The declaration states that:

1. The parties that have ratified Kyoto protocol strongly urge to others to do so in a timely manner (as nations declared in Johannesburg).

2. The IPCC's third assessment report confirms that the significant cuts in global emissions will be necessary to meet the convention's ultimate objective.

3. All the parties should continue to advance the implementation of their convention commitments and developed countries should demonstrate that they are taking the lead in modifying long-term emission trends.

4. The economic and social development and poverty eradication are the overriding priorities of developing countries.

5. The urgent action is needed to enable countries, in particular the least developed and small island countries to adapt the impacts of climate change.

6. The actions are required to develop cleaner, more efficient and affordable energy technologies, including fossil fuel and renewable energy technologies.

7. The actions are required, with a sense of urgency, to substantially increase the global share of renewable energy sources. The declaration was adopted by consensus. In the closing plenary session the European Union, Japan and Canada expressed their disappointment that it did not offer a clearer long-term vision. The EU said it would submit its own statement for the record. The developing countries and the United States expressed strong support for the declaration. The Nigeria thanked the United States for serving as a "constructive force" in the negotiations.

References

MAJOR SOURCE

The writer deeply studied and thoroughly investigated the Marxism, Leninism, Stalinism, Mao Tse Tung thought, modernization theory of China by Deng, Gorbachev's theory of Perestroika, Glasnost and democratization of Soviet Union and other communist trends of the world including the writings of Kim il Sung, Anwar Hoxa of Albania, the Model of International Democracy and the Nature-Human Centric Model under the leadership and guidance of late Sh. RP Saraf. The writer also deeply studied the capitalist school of thought, the historical process of corporate capitalism from its beginning to its present global corporate stage, the Indian Freedom Movement, multi-ethnic India, 1947's communal partition of India, the deep study of J&K dispute, South Asian Regions and having continuous touch with the global trends and fundamental threat of global warming/climate change and global dehumanization/human cum social inequality. The writer having long practice and experience of Marxist revolutionary movement and working class movement. The writer is also the founder of Democratic Movement (May 1994) and chief editor of view point of Democratic Movement. The writer after the deep study of nature and mankind (two basic factors of human society) and existential threat of global warming/climate change and global dehumanization/human inequality founded Nature-Mankind Friendly Global Movement on 2 October 2009 in a cadre conference at J&K part of India and also started the publication of Nature-Mankind Friendly Global View Point. The deep study with global perspective and nature-mankind friendly vision enables the writer to write this global historical book.

FROM THE INTERNET

1. www.voanews.com;
2. news.bbc.co.uk;
3. www.guardian.co.uk;
4. en.cop15.dk;
5. en.wikipedia.org/...2009_united_nations_climate_change_conference;
6. unfccc.int;en.wikipedia.org;
7. www.britannica.com;

8. timesofindia.indiatimes.com;
9. www.environmentalgraffiti.com;
10. environment.nationalgeographic.com;
11. www.realclimate.org;
12. www.global-greenhouse-warming.com/glacial-retreat.html;
13. www.greenpeace.org; unfccc.int

FROM PRINT AND ELECTRONIC MEDIA

1. Times of India (22/12/09); Articles: out of warming pan into fire and succumbing to pressure: Aam Admi sacrificed at US altar again by Surya P Sethi.

2. Marx.Engles. Marxism collected works of V. I. Lenin, volume prepared by Progressive Publishers, Moscow (1979); the Marxist Doctrine; page 13.

3. Programme of the nature-human centric people Movement; Feb. 2003; number 180; editor Sh. RP Saraf.

4. Nature-human centric viewpoint; documents of Jaipur conference; 25–27 November 2005; convened by Nepal-India national units of nature-human centric people movement.

5. Nature-Human centric viewpoint; July-Oct 2007, vol. 1, issue no. 19 ; an appeal submitted through UN Secretary General to 2007 UN world summit of the heads of national governments at Bali Indonesia on the question of responding to the challenge of Global warming or climate change; edited by Sh. R.P.Saraf.

6. Nature-human centric viewpoint; article; Bali roadmap prepares the way towards a deadly future by forging a fake unity in a politically, economically and culturally divided human world vol. 1, issue no. 21, January-February, 2008.

7. Nature-human centric viewpoint; edited by Sh. R.P.Saraf; five fatal challenges facing human community by 2050; vol.1, issue no. 22–25; March-October 2008.

8. Internationalist democratic viewpoint; July-August 2002, numbers 173–174 edited by Sh. RP Saraf; an appeal to 2002 UN world summit on sustainable development in Johannesburg; honourable heads of the government; choice is clear survival or extinction; there is no third alternative let this summit be a turning-point in securing our future.

9. Internationalist democratic viewpoint; September-October 2003, numbers 175-176; edited by Sh. R.P.Saraf; 2002 Johannesburg, UN Earth summit pushes human community towards a dangerous course by adopting a corporate agenda as the engine of sustainable development.

10. Nature-Human centric viewpoint; January-February 2005 vol. 1 issue no. 4; crisis of Agriculture; edited by Sh. R.P.Saraf.
11. Nature-Human centric viewpoint; April 2005, vol. 1, issue no. 7; Electric power today biggest source of energy as well as biggest source of pollution in Human society.
12. View point of Democratic Movement; Editor Babu Singh; May 1994-April 2004; number 1 to 5.
13. BBC London Hindi and English.
14. UNIPCC assessment reports on global warming and climate change
15. Plan B-3 and Plan B-4 by Lester R Brown, Founder Earth Policy Institute US.
16. The Sustainable Revolution by Peter J Macmanner.
17. The age of extremes-A History of the World,1914–1991 by Eric Hobsbawm.
18. The Grand Design by Stephen Hawking, A Great World Level Scientist of Austro-Physicist.
19. NIPCC Report "Climate Change Reconsidered 2009"
20. Open Society and Its Enemies by KR Popper.
21. The Wind of Change; The Arab World in Turmoil; By Ahmed Ali Ibrahim Sabeyse.

Glossary

1. *Anthropogenic adjective:* originating in human activity. (2) Dehumanize: verb: deprive of positive human qualities. Derivative: dehumanization (noun).

2. *Wavelength:* Noun; In Physics the distance between successive crests of a wave, especially as a distinctive feature of sound, light, radio waves, etc.

3. *Infra-red rays:* Portion of the electromagnetic spectrum that extends from the microwave range to the red end of the visible light range. Its wavelengths vary from about 0.7 to 1,000 micrometers. The trapping of infrared radiation by atmospheric gases is also the basis of the greenhouse effect.

4. *Absolute zero (0 Kelvin):* Temperature at which a thermodynamic system has the lowest energy, 0 Kelvin (K). It corresponds to -459.67°F (-273.15°C) and is the lowest possible temperature theoretically achievable by a system.

5. *Atmospheric gases:* Gases present in the atmosphere. The atmosphere of Earth is a layer of gases surrounding the planet Earth that is retained by Earth's gravity.

6. *Molecular collision:* Molecular collision is an isolated event in which two or more moving molecules (colliding bodies) exert relatively strong forces on each other for a relatively short time.

7. *Convection:* noun transference of mass or heat within a fluid caused by the tendency of warmer and less dense material to rise.

8. *Aerosols:* an aerosol is a suspension of fine solid particles or liquid droplets in a gas. Examples are smoke.

9. *Photosynthesis:* noun the process by which green plants uses sunlight to synthesize nutrients from carbon dioxide and water.

10. Thermal expansion: The dimensional changes exhibited by solids, liquids, and gases for changes in temperature while pressure is held constant.

11. *Temperature stratification:* Horizontal layers of differing densities produced in a lake by temperature changes.

12. *Ocean circulation:* Water current flow in a closed circular pattern within an ocean and the large-scale horizontal water motion within an ocean.

13. *Thermo-haline cycle:* The term thermo-haline circulation (THC) refers to the part of the large-scale ocean circulation that is driven by global density gradients created by surface heat and freshwater fluxes.

14. *Ice-albedo feedback* (or snow-albedo feedback): is a positive feedback climate process where a change in the area of snow-covered land, ice caps, glaciers or sea ice alters the albedo.

15. *Gigantic:* adjective of very great size or extent.

16. *Bio-diversity:* The variability among living organisms on the Earth, including the variability within and between species and within and between ecosystems.

17. *Bio-sequestration:* Bio-sequestration is the capture and storage of the atmospheric greenhouse gas carbon dioxide by an increased volume or quality of photosynthesis(through practices such as reforestation/preventing deforestation and genetic engineering respectively), enhanced soil carbon trapping in agriculture, as well as the use of algal bio-sequestration (see algae bioreactor) to absorb the constant stream of carbon dioxide emissions from coal-fired electricity generation.

18. *Slash-and-burn farming:* Slash and burn consists of cutting and burning of forests or woodlands to create fields for agriculture or pasture for livestock, or for a variety of other purposes. It is sometimes part of shifting cultivation agriculture, and of transhumance livestock herding.

19. *Deltas:* A usually triangular alluvial deposit at the mouth of a river.

20. *Adaptation:* The ability of a species to survive in a particular ecological niche, esp. because of alterations of form or behavior brought about through natural selection.

21. *Conjunction:* the state of being conjoined; union; association.

22. *Equivocal:* allowing the possibility of several different meanings, as a word or phrase, esp. with intent to deceive or misguide; susceptible of double interpretation; deliberately ambiguous.

23. *Embedded:* to fix into a surrounding mass.

24. *Guzzled:* to drink, or sometimes eat, greedily, frequently, or plentifully.

25. *Cosmic rays:* A stream of ionizing radiation of extraterrestrial origin, consisting chiefly of protons, alpha particles, and other atomic nuclei but including some high-energy electrons, that enters the atmosphere, collides with atomic nuclei and produces secondary radiation, principally pions, muons, electrons, and gamma rays.

26. *Discernible:* capable of being discerned; distinguishable.

27. *Interplay:* reciprocal relationship, action, or influence

28. *Indefensible:* incapable of being defended against criticism or denial; untenable.

29. *Genetically modified:* The scientific alteration of the structure of genetic material in a living organism. It involves the production and use of recombinant DNA and has been employed to create bacteria that synthesize insulin and other human proteins.

30. *Inorganic:* not characterized by vital processes; chemistry. Noting or pertaining to compounds that are not hydrocarbons or their derivatives.

31. *Consummated:* to bring to a state of perfection; fulfil.

32. *Dialectical materialism:* The Marxian interpretation of reality that views matter as the sole subject of change and all change as the product of a constant conflict between opposites arising from the internal contradictions inherent in all events, ideas, and movements.

33. *Mythical:* without foundation in fact; imaginary; fictitious.

34. *Repository:* A burial place; sepulchre.

35. *Totalitarian:* Of or pertaining to a centralized government that does not tolerate parties of differing opinion and that exercises dictatorial control over many aspects of life.

36. *Dogmatic:* of, pertaining to, or of the nature of a dogma or dogmas; doctrinal. 42 Hob-Nob: Hobnobbing with our social betters can be a hit-or-miss proposition, a fact that has an etymological justification.

37. *Confederal:* Of, relating to, or involving the activities of two or more nations.

38. *Jurisprudence:* The science or philosophy of law.

39. *Warmongering:* One who advocates or attempts to stir up war.

40. *Impoverishment:* to make poor in quality, productiveness, etc.; exhaust the strength or richness.

41. *Shun:* to keep away from (a place, person, object, etc.), from motives of dislike, caution, etc.; take pains to avoid.

42. *Extravagant:* spending much more than is necessary or wise; wasteful.

43. *Lavish:* expended, bestowed, or occurring in profusion.

44. *Catastrophic:* a sudden and widespread disaster.

45. *Mitigation:* Mitigation is the effort to reduce loss of life and property by lessening the impact of disasters.

Index